Essentials of
Control Techniques
and Theory

Essentials of Control Techniques and Theory

John Billingsley

CRC Press
Taylor & Francis Group
Boca Raton   London   New York

CRC Press is an imprint of the
Taylor & Francis Group, an **informa** business

CRC Press
Taylor & Francis Group
6000 Broken Sound Parkway NW, Suite 300
Boca Raton, FL 33487-2742

First issued in paperback 2017

© 2010 by Taylor and Francis Group, LLC
CRC Press is an imprint of Taylor & Francis Group, an Informa business

No claim to original U.S. Government works

ISBN 13: 978-1-138-11631-3 (pbk)
ISBN 13: 978-1-4200-9123-6 (hbk)

Library of Congress Cataloging-in-Publication Data

Billingsley, J. (John)
   Essentials of control techniques and theory / John Billingsley.
       p. cm.
   Includes index.
   ISBN 978-1-4200-9123-6 (hardcover : alk. paper)
   1. Automatic control. 2. Control theory. I. Title.

TJ223.M53B544 2010
629.8--dc22

2009034834

Visit the Taylor & Francis Web site at
http://www.taylorandfrancis.com

and the CRC Press Web site at
http://www.crcpress.com

Contents

SECTION II ESSENTIALS OF CONTROL THEORY—WHAT YOU OUGHT TO KNOW

Preface

I am always suspicious of a textbook that promises that a subject can be "made easy." Control theory is not an easy subject, but it is a fascinating one. It embraces every phenomenon that is described by its variation with time, from the trajectory of a projectile to the vagaries of the stock exchange. Its principles are as essential to the ecologist and the historian as they are to the engineer.

All too many students regard control theory as a backpack of party tricks for performing in examinations. "Learn how to plot a root locus, and question three is easy." Frequency domain and time domain methods are often pitted against each other as alternatives, and somehow the spirit of the subject falls between the cracks. Control theory is a story with a plot. State equations and transfer functions all lead back to the same point, to a representation of the actual system that they have been put together to describe.

The subject is certainly not one that can be made easy, but perhaps the early, milder chapters will give the student an appetite for the tougher meat that follows. They should also suggest control solutions for the practicing engineer. The intention of the book is to explain and convince rather than to drown the reader in detail. I would like to think that the progressive nature of the mathematics could open up the early material to students at school level of physics and computing—but maybe that is hoping too much.

The computer certainly plays a large part in appreciating the material.

With the aid of a few lines of software and a trusty PC, the reader can simulate dynamic systems in real time. Other programs, small enough to type into the machine in a few minutes, give access to methods of on-screen graphical analysis methods including Bode, Nyquist, Nichols, and Root Locus in both s- and z-planes. Indeed, using the Alt-PrintScreen command to dump the display to the clipboard, many of the illustrations were produced from the programs that are listed here and on the book's Web site.

There are many people to whom I owe thanks for this book. First, I must mention Professor John Coales, who guided my research in Cambridge so many years ago. I am indebted to many colleagues over the years, both in industry and academe. I probably learned most from those with whom I disagreed most strongly!

My wife, Rosalind, has kept up a supply of late-night coffee and encouragement while I have pounded the text into a laptop. The illustrations were all drawn and a host of errors have been corrected. When you spot the slips that I missed, please email me so that I can put a list of errata on the book's Web site: http://www. esscont.com. There you will find links for my email, software simulation examples, and a link to the publisher's site.

Now it is all up to the publishers—and to you, the readers!

Author

John Billingsley graduated in mathematics and in electrical engineering from Cambridge University in 1960. After working for four years in the aircraft industry on autopilot design, he returned to Cambridge and gained a PhD in control theory in 1968.

He led research teams at Cambridge University developing early "mechatronic" systems including a laser phototypesetting system that was the precursor of the laser printer and "acoustic telescope" that enabled sound source distributions to be visualized (this was used in the development of jet engines with reduced noise).

He moved to Portsmouth Polytechnic in 1976, where he founded the Robotics Research Group. The results of the Walking Robot unit led to the foundation of Portech Ltd., which for many years supplied systems to the nuclear industry for inspection and repair of containment vessels. Other units in the Robotics Research Group have had substantial funding for research in quality control and in the integration of manufacturing systems with the aid of transputers.

In April 1992 he took up a Chair of Engineering at the University of Southern Queensland (USQ) in Toowoomba. His primary concern is mechatronics research and he is Director of Technology Research of the National Centre for Engineering in Agriculture (NCEA).

Three prototypes of new wall-climbing robots have been completed at USQ, while research on a fourth included development of a novel proportional pneumatic valve. Robug 4 has been acquired for further research into legged robots.

A substantial project in the NCEA received Cotton Research funding and concerned the guidance of a tractor by machine vision for very accurate following of rows of crop. Prototypes of the system went on trial on farms in Queensland, New South Wales, and the United States for several years. Novel techniques are being exploited in a further commercial project. Other computer-vision projects have included an automatic system for the grading of broccoli heads, systems for discriminating between animal species for controlling access to water, systems for precision counting and location of macadamia nuts for varietal trials, and several other systems for assessing produce quality.

Dr. Billingsley has taken a close interest in the presentation of engineering challenges to young engineers over many years. He promoted the Micromouse robot maze contest around the world from 1980 to the mid-1990s.

He has contrived machines that have been exhibited in the "Palais de la Decouverte" in Paris, in the "Exploratorium" at San Francisco and in the Institute of Contemporary Arts in London, hands-on experiments to stimulate an interest in control. Several robots resulting from projects with which Dr. Billingsley was associated are now on show in the Powerhouse Museum, Sydney.

Dr. Billingsley is the international chairman of an annual conference series on "Mechatronics and Machine Vision in Practice" that is now in its sixteenth year.

He was awarded an Erskine Fellowship by the University of Canterbury, New Zealand, where he spent February and March 2003.

In December 2006 he received an achievement medal from the Institution of Engineering and Technology, London.

His published books include: *Essentials of Mechatronics*, John Wiley & Sons, June 2006; *Controlling with Computers*, McGraw-Hill, January 1989; *DIY Robotics and Sensors on the Commodore Computer*, 1984, also translated into German: *Automaten und Sensoren zum selberbauen*, Commodore, 1984, and into Spanish: *Robotica y sensores para el commodoro-proyectos practicos para aplicaciones de control*, Gustavo Gili, 1986; *DIY Robotics and Sensors with the BBC Computer*, 1983.

John Billingsley has also edited half a dozen volumes of conference proceedings, published in book form.

ESSENTIALS OF CONTROL TECHNIQUES—WHAT YOU NEED TO KNOW

1

Chapter 1

Introduction: Control in a Nutshell; History, Theory, Art, and Practice

There are two faces of automatic control. First, there is the theory that is required to support the art of designing a working controller. Then there is further, and to some extent, different theory that is required to convince a client, employer, or examiner of one's expertise.

You will find both covered here, carefully arranged to separate the essentials from the ornamental. But perhaps that is too dismissive of the mathematics that can help us to understand the concepts that underpin the controller's effects. And if you write up your control project, the power of mathematical terminology can elevate a report on simple pragmatic control to the status of a journal paper.

1.1 The Origins of Control

We can find early examples of control from long before the age of "technology." To flush a toilet, it was once necessary to tip a bucket of water into the pan—and then walk to the pump to refill the bucket. Then a piped water supply meant that a tap could be turned to fill a cistern—but you had to remember to turn off the tap. But today everyone expects a ball-shaped float on an arm inside the cistern to turn off the water automatically—you can flush and forget.

There was rather more engineering in the technology that turned a windmill to face the wind. These were not the "Southern Cross" iron-bladed machines that can be seen pumping water from bores across Australia, but the traditional windmills for which Holland is so famous. They were too big and heavy to be rotated by a simple weather vane, so when the millers tired of lugging them round by hand they added a small secondary rotor to do the job. This was mounted at right angles to the main rotor, to catch any crosswind. As this rotated it used substantial gearing to crank the whole mill round in the necessary direction to face the wind.

Although today we could easily simulate either of these systems, it is most unlikely that any theory was used in their original design.

While thinking of windmills, we can see that there is often a simple way to get around a technological problem. When the wind blows onto the shore in the daytime, or off the shore at night, the Greeks have an answer that does not involve turning the mill around at all. The rotor consists of eight triangular sails flying from crossed poles, each rather like the sail of a wind-surfer. Just as in the case of a wind-surfer, when the sail catches the wind from the opposite side the pole is still propelled forward in the same direction.

Even more significant is the technique used to keep a railway carriage on the rails. Unlike a toy train set, the flanges on the wheels should only ever touch the rails in a crisis. The control is actually achieved by tapering the wheels, as shown in Figure 1.1. Each pair of wheels is linked by a solid axle, so that the wheels turn in unison.

Now suppose that the wheels are displaced to the right. The right-hand wheel now rolls forward on a larger diameter than the left one. The right-hand wheel travels a little faster than the left one and the axle turns to the left. Soon it is rolling to the left and the error is corrected. But as we will soon see, the story is more complicated than that. As just described, the axle would "shimmy," oscillating from side to side. In practice, axles are mounted in pairs to form a "bogey." The result is a control system that behaves as needed without a trace of electronics.

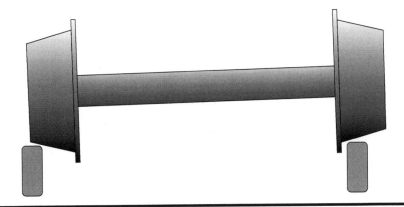

Figure 1.1 A pair of railway wheels.

1.2 Early Days of Feedback

When the early transatlantic cables were laid, amplifiers had to be submerged in mid-ocean. It was important to match their "gain" or amplification factor to the loss of the cable between repeaters. Unfortunately, the thermionic valves used in the amplifiers could vary greatly in their individual gains and that gain would change with time. The concept of feedback came to the rescue (Figure 1.2).

A proportion of the output signal was subtracted from the input. So how does this help?

Suppose that the gain of the valve stage is A. Then the input voltage to this stage must be $1/A$ times the output voltage. Now let us also subtract k times the output from the overall input. This input must now be greater by kv_{out}. So the input is given by:

$$v_{in} = (1/A + k)v_{out}$$

and the gain is given by:

$$v_{out}/v_{in} = \frac{1}{k + 1/A}$$

$$= \frac{1/k}{1 + 1/Ak}.$$

So what does this mean? If A is huge, the gain of the amplifier will be $1/k$. But when A is merely "big," the gain fails short of expectations by factor of $1 + 1/(Ak)$. We have exchanged a large "open loop" gain for a smaller one of a much more certain value. The greater the value of the "loop gain" Ak, the smaller is the uncertainty.

But feedback is not without its problems. Our desire to make the loop gain very large hits the problem that the output does not change instantaneously with the input. All too often a *phase shift* will impose a limit on the loop gain we can apply before instability occurs. Just like a badly adjusted public-address microphone, the system will start to "squeal."

So the electronic engineers built up a large body of experience concerning the analysis and adjustment of linear feedback systems. To test the gain of an amplifier,

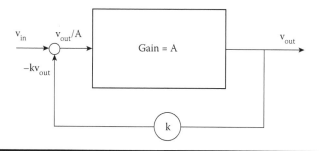

Figure 1.2 Effect of feedback on gain.

a small sinusoidal "whistle" from an *oscillator* was applied to the input. A variable *attenuator* could reduce the size of an oscillator's one-volt signal by a factor of, say, 100. If the output was then found to be restored to one volt, the *gain* was seen to be 100. (As the amplifier said to the attenuator, "Your loss is my gain." Apologies!)

As the frequency of the oscillator was varied, the gain of the amplifier was seen to change. At high frequencies it would *roll off* at a rate measured in *decibels per octave*—the oscillators had musical origins and levels were related to "loudness."

Some formal theory was needed to validate the rules of thumb that surrounded these plots of gain against frequency. The engineers based their analysis on complex numbers. Soon they had embroidered their methods with Laplace transforms and a wealth of arcane graphical methods, Bode diagrams, Nyquist diagrams, Nicholls charts and root locus, to name but a few. Not surprisingly, this approach was termed the *frequency domain*.

When the control engineers were faced with problems like simple position control or the design of autopilots, they had similar reasons for desiring large loop gains. They hit stability problems in just the same way. So they "borrowed" the frequency-domain theory lock, stock, and barrel.

Unfortunately, few real control systems are truly linear. Motors have limits on how hard they can be driven, for a start. If a passenger aircraft banks at more than an angle of 30°, there will probably be complaints if not screams from the passengers. Methods were needed for *simulating* the systems, for finding how they would respond as a function of time.

1.3 The Origins of Simulation

The heart of a simulation is the *integrator*. Of course we need some differential equations to start with. If the output of an integrator is x, then its input is dx/dt. By cascading integrators we can construct a differential equation of virtually any order. But where can we find an integrator?

In the Second World War, bomb-aiming computers used the "ball and plate" integrator. A disk rotated at constant speed. A ball-bearing was located between the plate and a roller, being moved from side to side as shown in Figure 1.3. When the ball is held at the center of the plate, it does not move, so neither does the roller. If it is moved outward along the roller, it will pick up a rotation proportional to the distance from the center, so the roller will turn at a proportional speed. We have an integrator!

But for precision simulation, a "no-moving-parts" electronic system was needed. By applying feedback around an amplifier using a capacitor, we have feedback current proportional to the rate-of-change of the output. This cancels out the current from the input and once again we have an integrator.

Unfortunately, in the days of valves the amplifiers were not easy to make. The output had to vary to both positive and negative voltages, for a very small change

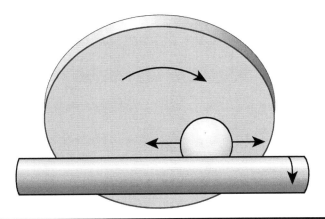

Figure 1.3   Ball-and-plate integrator.

in an input voltage near zero. Conventional amplifiers were *AC coupled*, used for amplifying speech or music. These new amplifiers had to give a constant DC output for a constant input. In an effort to compensate for the drift of the valves, some were *chopper stabilized*.

But in the early 1960s, the newfangled transistor came to the rescue. By then, both PNP and NPN versions were available, allowing the design of circuits where the output was pulled up or down symmetrically.

Within a few years, the manufacturers had started to make "chips" with complete circuits on them and an early example was the *operational amplifier*, just the thing the simulator needs. These have become increasingly more sophisticated, while their price has dropped to a few cents.

Just when perfection was in sight for the analog computer (or simulator), the digital computer moved in as a rival. Rather than having to patch circuitry together, the control engineer only needs to write a few lines of software to guarantee a simulation with no drift, no uncertainty of gain or time-constants, and an output that can produce a plot only limited by the engineer's imagination.

While the followers of the frequency-domain methods concern themselves with *transfer functions*, simulation requires the use of *state equations*. You just cannot escape mathematics!

1.4 Discrete Time

Simulation has changed the whole way we view control theory. When analog integrators were connected to simulate a system, the list of inputs to each integrator came to be viewed as a set of state equations, with the output of the integrators representing *state variables*.

Computer simulation and discrete time control go hand in hand together. At each iteration of the simulation, new values are calculated for the state variables in terms of their previous values. New input values are set that remain constant over the interval until the next iteration. We might be cautious at first, defining the interval to be so short that the calculation approximates to integration. But by examining the exact way that one set of state variables leads to the next, we can make the interval arbitrarily long.

Although discrete time theory is usually regarded as a more advanced topic than the frequency domain, it is in fact very much simpler. Whereas the frequency domain is filled with complex exponentials, discrete time solutions involve powers of a simple variable—though this variable may be complex, too.

By way of an example, if interest on your bank overdraft causes it to double after m months, then after a further m months it will double again. After n periods of m months, it will have been multiplied by 2^n. We have a simple solution for calculating its values at these discrete intervals of time. (Paying it off quickly would be a good idea.)

To calculate the response of a system and to assess the effect of discrete time feedback, a useful tool is the *z-transform*. This is usually explained in terms of the *Laplace transform*, but its concept is much simpler.

In calculating a state variable x from its previous value and the input u, we might have a line of code of the form:

```
x = ax + bu;
```

Of course this is not an equation. The x on the left is the new value while that on the right is the old value. But we can turn it into an equation by introducing an *operator* that means *next*. We denote this operator as z.

So now

$$zx = ax + bu$$

or

$$x = \frac{bu}{z - a}.$$

In the later chapters all the mysteries will be revealed, but before then we will explore the more conventional approaches.

You might already have noticed that I prefer to use the mathematician's "we" rather than the more cumbersome passive. Please imagine that we are sitting shoulder to shoulder, together pondering the abstruse equations that we inevitably have to deal with.

Chapter 2

Modeling Time

2.1 Introduction

In every control problem, time is involved in some way. It might appear in an obvious way, relating the height at each instant of a spacecraft, in a more subtle way as a list of readings taken once per week, or unexpectedly as a feedback amplifier bursts into oscillation.

Occasionally, time may be involved as an explicit function, such as the height of the tide at four o'clock, but more often its involvement is through differential or difference equations, linking the system behavior from one moment to the next. This is best seen with the example of Figure 2.1.

2.2 A Simple System

Figure 2.1 shows a cup of coffee that has just been made. It is rather too hot at the moment, at 80°C. If left for some hours it would cool down to room temperature at 20°C, but just how fast is it going to cool, and when will it be at 60°C?

The rate of fall in temperature will be proportional to the rate of loss of heat. It is a reasonable assumption that the rate of loss of heat is proportional to the temperature above ambient, so we see that if we write T for temperature,

$$\frac{dT}{dt} = k(T - T_{\text{ambient}}).$$

Figure 2.1 A cooling cup of coffee.

Figure 2.2 A leaking water butt.

If we can determine the value of the constant k, perhaps by a simple experiment, and then the equation can be solved for any particular initial temperature—the form of the solution comes later.

Equations of this sort apply to a vast range of situations. A rainwater butt has a small leak at the bottom as shown in Figure 2.2. The rate of leakage is proportional to the depth, H, and so:

$$\frac{dH}{dt} = -kH.$$

The water will leak out until eventually the butt is empty. But suppose now that there is a steady flow *into* the butt, sufficient to raise the level (without leak) at a speed u. Then the equation now becomes:

$$\frac{dH}{dt} = -kH + u.$$

At what level will the water settle down now? When it has reached a steady level, no matter how long it takes, the rate of change of depth will have fallen to zero, so $dH/dt = 0$. It is not hard to see that $-kH + u$ must also be zero, and so $H = u/k$.

Now if we really want to know the depth as a function of time, a mathematical formula can be found for the solution. But let us try another approach first, simulation.

2.3 Simulation

With a very little effort, we can construct a computer program that will imitate the behavior of the water level. If the depth right now is H, then we have already described the rate of change of depth dH/dt as $(-k\,H + u)$. In a short time dt, the depth will have changed by the rate of change multiplied by the interval

$$(-k\,H + u)dt.$$

To get the new value of H we add this to the old value. In programming terms we can write:

```
H = H + (-k*H + u)*dt
```

Although it might look like an equation, this is an *assignment statement* that gives a new value to H. It will work as it stands as a line of Basic. For C or Java add a semicolon. We can add another line to calculate the time,

```
t = t + dt
```

To make a simulation, we must wrap this in a loop. In one of the dialects of Basic this could be:

```
while (t < tmax)
    H = H + (-k*H + u)*dt
    t = t + dt
```

```
wend
```

or for JavaScript or C:

```
while (t < tmax){
  H = H + (-k*H + u)*dt;
  t = t + dt;
}
```

Now the code has to be "topped" to set initial values for H, t, and u, and it will calculate H until t reaches the time *tmax*. But although the computer might "know the answer," we have not yet added any output statement to let it tell us.

Although a "print" statement would reveal the answer as a list of numbers, we would really prefer to see a graph. We would also like to be able to change the input as the simulation goes on. So what computing environment can we use?

2.4 Choosing a Computing Platform

Until a few years ago, the "language of choice" would have been Quick Basic, or the Microsoft version QBasic that was bundled with Windows (although it ran under DOS). Simple PSET and LINE commands are all that is needed for graph plotting, while a printer port can be given direct output commands for driving motor amplifiers or reading data from interface chips.

With even better graphics, but less interfacing versatility is Visual Basic. Version 6.0 of Visual Studio was easy to use, based on the concept of forms on which graphs could be plotted with very similar syntax to QBasic. There are two problems with Visual Basic 6.0. Firstly, it probably cost more than this book, secondly, it has been superseded and might be unobtainable.

Microsoft is now offering a free download of .Net versions of Visual C++ and Visual Basic at www.microsoft.com/express. Unfortunately, the display of graphs in these new versions is no trivial matter.

So we need an environment that is likely to endure several generations of software updates by the system vendors. The clear choice is the browser environment, using the power of JavaScript.

Every browser supports a scripting language, usually employed for animating menu items and handling other housekeeping. The JavaScript language is very much like C in appearance. Even better, it contains a command "eval" that enables you to type your code as text in a window on the web page and then see it executed in real time.

By accessing this book's web page at *www.esscont.com*, you can download an example page with an "applet." This is a very simple piece of code that acts as a hook into the graphics facilities of Java. The web page acts as a "wrapper" to let you write simple routines in the code window to do all you need for graph plotting and more. A complete listing is given in Chapter 3; it is not very complicated.

But there's more.

By putting control buttons on the web page, you can interact with the simulation in real time, in a way that would be difficult with an expensive proprietary package.

By way of an example, let us wrap the simulation in a minimum of code to plot a graph in a web page.

Firstly, we have:

```
<head>
<script language="javaScript">
```

Then we put some "housekeeping" and the simulation inside a function, so that we can call it after the loading page has settled down.

```
function simulate() {
    document.Graph.init();
    g = document.Graph.g;
    document.Graph.setRGB(0,0,0);
```

These lines initialize the applet, then copy all its properties to a local variable "g." The final line uses one of these properties to set the pen color to black.

Now, at last we can start the simulation, first with the definition of the variables we will use.

```
    var k = .5;
    var dt = .01;
    var t=0;

    var x=0; //Initial level 0 to 40
    var u=20; //Input flow 0 to 20

    while (t<10) {
      oldx=x;
      oldt=t;

      x = x + (-k*x + u)*dt; //This is the simulation
      t = t + dt

      g.drawLine(50*oldt,400-10*oldx,50*t,400-10*x);
    //We have to do the scaling and the applet's y increases
    downwards
      }
}

</script>
</head>
```

Now we design a web page with a gray background and the Graph applet centered

```
<body style="background-color: rgb(200, 200, 200);">
<center>
<applet code="Graph.class" name="Graph" height="400"
width="640">
</applet></center>
```

Finally, we set up the execution of the simulation to start 200 milliseconds after the page has loaded and tidy up the end of the file.

```
<script>
  setTimeout('simulate()',200);
</script>
</body>
</html>
```

Open Notepad or Wordpad and type in the code above. Save it as a text file with title *sim.htm*, in the same folder as a copy of *graph.class*. Do not close the editor.

(The code and applet can be found on the book's website at www.esscont. com/2/sim.htm.)

Q 2.4.1

Open the file with your browser, any should do, and soon after the page has loaded you will see a graph appear. Sketch it.

Q 2.4.2

Now edit the code to give initial conditions of $x = 40$ and $u = 0$. Save the file and reload the browser web page. What do you see?

Q 2.4.3

Edit the file again to set the initial $x = 0$ and $u = 10$. Save it again and reload the web page. What do you see this time?

Q 2.4.4

Edit the file again to set $dt = 1$. What do you see this time?

Q 2.4.5

Now try $dt = 2$.

Q 2.4.6

Finally try *dt* = 4.

2.5 An Alternative Platform

A new graphics environment is being developed under HTML5. At present it is only available in Mozilla browsers such as Firefox and Seamonkey. There are some inconsistencies in the definition, so that changes might be expected before long. Nevertheless it enables graphics to be displayed without the need to employ an applet.

For the rest of the book, the applet approach will be followed. However, if you wish to see more of the use of "canvas," do the following:

First download and install Firefox—it is cost free.

Enter the following code into Notepad, save it as *simcanvas.htm* and open it with Firefox. You should see the response of Figure 2.3.

```
<html>
<head>

<script type="application/x-javascript">
function simulate() {
  var canvas = document.getElementById("canvas");
  var ctx = canvas.getContext("2d");
  ctx.strokeStyle = "blue";

  var k = .5; //This part is the same as before
  var dt = .01; //Edit to try various steplengths
  var t=0;
```

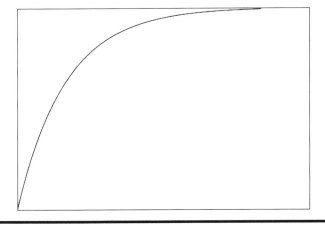

Figure 2.3   Screen grab of *simcanvas.htm*.

```
x=0; //Initial level 0 to 40
u=20; //Input flow 0 to 20

ctx.beginPath(); //Get ready to plot
ctx.moveTo(50*t,400-10*x);

while (t<10) {
   x = x + (-k*x + u) * dt; //This is the simulation
   t = t + dt
   ctx.lineTo(50*t,400-10*x);
}
ctx.stroke(); //render the plotted path
}
</script>
<style type="text/css">
   canvas {border: 2px solid black; background: white;}
</style>
</head>

<body style="background-color: rgb(200, 200, 200);"
onLoad="simulate();">
<center>
   <canvas id="canvas" width="600" height="400" "></canvas>
</center>
</body>
</html>
```

2.6 Solving the First Order Equation

You should have noticed that the result of the simulation varies if you change the value of *dt*. For small step lengths the change might be negligible, but for values approaching one second the approximation deteriorates rapidly. The calculation is based on the assumption that the rate of change remains constant throughout the step interval, but if the interval is not small the errors will start to build up.

A practical solution is to halve the value of *dt* and repeat the simulation. If no change can be seen, the interval is short enough. Another strategy might be to use a much more complicated integration process, such as Runge–Kutta, but it is preferable to keep the code as simple as possible.

There is a way to perform an accurate computer simulation with no limit on step size, provided that the input does not change between steps.

We will consider the formal solution of the simple example. The treatment here may seem over elaborate, but later on we will apply the same methods to more demanding systems.

By using the variable x instead of H or T_{coffee}, we can put both of the simple examples into the same form:

$$\dot{x} = ax + bu. \qquad (2.1)$$

Here, the dot over the x means *rate of change of x, dx/dt*. a and b are constants which describe the system, and u is an input, which might simply be a constant in the same way that T_{ambient} is in the coffee example.

Rearranging, we see that:

$$\dot{x} - ax = bu. \qquad (2.2)$$

Since we have a mixture of x and dx/dt, we cannot simply integrate the equation. We must somehow find a function of x and time which when differentiated will give us both terms on the left of the equation.

Let us consider

$$\frac{d}{dt}(x\,f(t))$$

where $f(t)$ is some function of time which has derivative $f'(t)$. When we differentiate by parts we see that:

$$\frac{d}{dt}(x\,f(t)) = \dot{x}\,f(t) + x\,f'(t). \qquad (2.3)$$

If we multiply Equation 2.2 through by $f(t)$, we get:

$$\dot{x}\,f(t) - a\,f(t)\,x = b\,u\,fs(t). \qquad (2.4)$$

Now if we can find $f(t)$ such that,

$$f'(t) = -a\,f(t)$$

then Equation 2.4 will become

$$\frac{d}{dt}(x\,f(t)) = b\,u\,f(t)$$

and we will have something that we can integrate.

The solution to our mystery function is

$$f(t) = e^{-at}$$

and our equation becomes

$$\frac{d}{dt}(xe^{-at}) = bue^{-at}. \tag{2.5}$$

When we integrate this between the limits 0 and t, we find:

$$\left[xe^{-at}\right]_0^t = \int_0^t bue^{-at}\,dt.$$

If u remains constant throughout the interval, we can take it outside the integral. We get:

$$x(t) \cdot e^{-at} - x(0) \cdot 1 = u\left[b\frac{-1}{a}e^{-at}\right]_0^t,$$

$$\text{so } x(t)e^{-at} = x(0) + u\frac{b}{a}(1 - e^{-at}). \tag{2.6}$$

We can multiply through by e^{-at} to get

$$x(t) = x(0)e^{at} + u\frac{b}{a}(e^{at} - 1). \tag{2.7}$$

This expresses the new value of x in terms of its initial value and the input in the form:

$$x(t) = gx(0) + hu(0)$$

where

$$g = e^{at}$$

and

$$h = b\frac{e^{at} - 1}{a}.$$

Now we can see how this relates to simulation with a long time-step. For steps of a constant width *T*, we can write:

$$x((n+1)T) = gx(nT) + hu(nT)$$

and the computer code to calculate the next value is simply

```
x = g x + h u
```

There is no "small dt" approximation, the expression is exact when the constants *g* and *h* have been calculated from the value of *T*.

The simulation program of the last section can now be made precise. We replace *a* with the value −*k* and *b* with value 1, so that for step length *dt* we have:

```
var k = .5;
var dt = 1;
var t=0;

var g = Math.exp(-k * dt);
var h = (1-g) / k;

var x=0; //Initial level 0 to 40
var u=20; //Input flow 0 to 20

while (t<10) {
  oldx=x;
  oldt=t;

  x = g*x +h*u; //This is the simulation
  t = t + dt

  g.drawLine(50*oldt,400-10*oldx,50*t,400-10*x);
                //because the applet's y increases downwards
}
```

The magic is all in that one line that calculates a new value for *x*. It almost looks too easy. But can the method cope with higher order systems?

2.7 A Second Order Problem

A servomotor drives a robot axis to position *x*. The speed of the axis is *v*. The acceleration is proportional to the drive current *u*; at present there is no damping. Can we model the system to deduce its performance? (Figure 2.4).

In Section 2.3, we had an equation for the rate of change of a variable that described the system. By repeatedly adding the small change over a short interval,

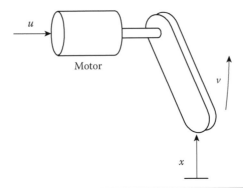

Figure 2.4 Position control example.

we were able to track the progress of the system against the time. Let us try the same approach here.

One of the variables describing the system is the position *x*, so we look for an equation for *dx/dt*. When we spot it, it is simply:

$$\frac{d}{dt}x = v.$$

Clearly, we have found a second "state variable," *v*, and we must look for an expression for its rate of change. But we have been told that this acceleration is proportional to the input, *u*.

So our second equation is:

$$\frac{d}{dt}v = b\,u.$$

Now we have two differential equations and their right-hand-sides only contain variables that we "know," so we can set up two lines of code

```
x = x + v*dt
v = v + b*u*dt
```

and we have only to "top-and-tail" them and wrap them in a loop to make a simulation.

If we modify the same code that we used previously, we have:

```
<head>
<script language="javaScript">

function simulate() {
    document.Graph.init();
    g=document.Graph.g;
    document.Graph.setRGB(0,0,0);
```

```
    var b=1;
    var dt=.01;
    var t=0;

    var x=-10; //Initial position
    var v=0;
    var u=20; //Input motor current

    while (t<10) {
      oldx=x;
      oldt=t;

      x = x + v*dt; //This is the simulation
      v = v + b*u*dt;
      t = t + dt

      g.drawLine(50*oldt,400-10*oldx,50*t,400-10*x);
    }
}

</script>
</head>

<body style="background-color: rgb(200, 200, 200);">
<center>
<applet code="Graph.class" name="Graph" height="400"
width="640">
</applet></center>

<script>
   setTimeout('simulate()',2000);
</script>
</body>
</html>
```

(Saved as www.esscont.com/2/sim2.htm on the book's website.)

As it stands, the constant input rapidly drives the position off the screen as shown in Figure 2.5. The simulation only becomes interesting when we attempt some position control. Let us make the drive proportional to the position error by adding a line

```
u = 4*(xtarget-x);
```

before the " x =" line and

```
var xtarget=0;
```

among the definitions.
   (Saved as www.esscont.com/2/sim3.htm)

Figure 2.5   Screen grab of *sim2.htm*.

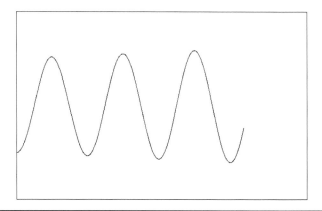

Figure 2.6   Screen grab of *sim3.htm*.

We discover from Figure 2.6 that feeding back position alone just results in oscillation. The oscillations appear to grow because of the simulation error caused by the finite value of *dt*. Try reducing it, say, to 0.001.

To get stability we need to add some damping in the form of a velocity term. In fact we can try for a rapid response without overshoot. Try

```
u = 4*(xtarget-x) - 2*v;
```

then adjust the coefficient of *v* to get the "best" response.
(See www.esscont.com/2/sim4.htm)

Q 2.7.1

What is the smallest value that will avoid an overshoot?

Q 2.7.2

What is the result of setting the *v* coefficient to 20?

Q 2.7.3

What is the result if you set

```
u = 64*(xtarget-x)-10*v;
```

Q 2.7.4
Now try

```
u = 100*(xtarget-x)-20*v;
```

It looks as though we can get an arbitrarily fast response with very little effort. We will soon be disillusioned!

2.8 Matrix State Equations

Clearly we have succeeded in finding a way to simulate the second order problem, and there seems no reason why the same approach should not work for third, fourth, fifth... How can the approach be formalized?

Firstly, we must find a set of variables that describe the present state of the system—in this case *x* and *v*. These we will glorify with the title *State Variables*. They must all have derivatives that can be expressed as combinations of just the variables and the input (or inputs). Then we will have a set of equations, with no unknown factors, which express the change of each variable from instant to instant. If there should happen to be some unknown term, then we have clearly left out one of the state variables, and we must hunt for its derivative to work it in as an extra equation.

In the present example, the equations can be laid out as:

$$\frac{dx}{dt} = 1v$$

$$\frac{dv}{dt} = bu$$

which can be written in matrix form as

$$\begin{bmatrix} \dot{x} \\ \dot{v} \end{bmatrix} = \begin{bmatrix} 0 & 1 \\ 0 & 0 \end{bmatrix} \begin{bmatrix} x \\ v \end{bmatrix} + \begin{bmatrix} 0 \\ b \end{bmatrix} u.$$

Admittedly, the elaboration appears somewhat useless in this simple example, but we will soon find that system after system falls into the matrix mould of:

$$\dot{x} = Ax + Bu$$

where **x** is a vector consisting of all the state variables.

2.9 Analog Simulation

It is ironic that analog simulation "went out of fashion" just as the solid-state operational amplifier was perfected. Previously, the integrators had involved a variety of mechanical, hydraulic, and pneumatic contraptions, followed by an assortment of electronics based on magnetic amplifiers or thermionic valves. Valve amplifiers were common even in the late 1960s, and required elaborate stabilization to overcome their drift. Power consumption was higher, and air-conditioning essential.

Soon an operational amplifier was available on a single chip, then four to a chip at a price of a few cents. But by then it was deemed easier and more accurate to simulate a system on a digital computer. The cost of analog computing had become that of achieving tolerances of 0.01% for resistors and capacitors, and of constructing and maintaining the large precision patch-boards on which each problem was set up.

In the laboratory, the analog computer still has its uses. Leave out the patch-board, and solder up a simple problem directly. Forget the 0.1% components—the parameters of the system being modeled are probably not known to better than a percent or two, anyway. Add a potentiometer or two to set up feedback gains, and a lot of valuable experience can be acquired. Take the problem of the previous section, for example.

An analog integrator is made from an operational amplifier by connecting the non-inverting input to a common rail, analog ground, while the amplifier supplies are at +12 volts and −12 volts relative to that rail (more or less). The inverting input is now regarded as a *summing junction*. A feedback capacitor of, say, 10 microfarads connects the output to this junction, while inputs are each connected via their own 100 kilohm resistor. Such an integrator will have a *time constant* of one second. If +1 volt is applied at one input, the output will change by one volt (negatively) in one second.

To produce an output that will change in the same sense as that of the input, we must follow this integrator with an invertor, a circuit that will give −1 volt output for +1 volt input. Another operational amplifier is used, also with the non-inverting input grounded. The feedback to the summing junction now takes the form of a 100 kilohm resistor, and the single input is connected through another 100 kilohm resistor as shown in Figure 2.7.

The gain of a virtual-earth amplifier is deduced by assuming that the feedback succeeds always in maintaining the summing junction very close to the analog ground, and by assuming that no input current is taken by the chip itself. The

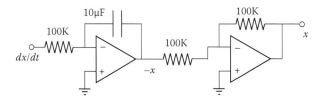

Figure 2.7   Construction of an integrator.

feedback current of an integrator is then $C\,dv/dt$, where v is the output voltage. The input current is v_{in}/R, and so:

$$C\frac{dV}{dt} + \frac{1}{R}v_{in} = 0.$$

Now if the integrator's 10 microfarad feedback is reduced to 1 microfarad, its time constant will only be 0.1 second. If instead the resistor R is doubled to 200 kilohms the time constant will also be doubled. Various gains can be achieved by selecting appropriate input resistors.

Now in our position control example we have two "state variables" x and v, governed by:

$$\frac{dx}{dt} = v$$

and

$$\frac{dv}{dt} = bu.$$

If we set up two integrators, designating the outputs as x and v, and if we connect the output of the v integrator to the input of the x integrator, and the input signal u to the input of the v integrator through an appropriate resistor, then the simulation of Figure 2.8 is achieved.

As in the case of the digital simulation, the exercise only becomes interesting when we add some feedback.

As in the digital simulation, we want to make

$$u = f(x_{demand} - x) - d\,v$$

where x_{demand} is a target position, f is the feedback gain, and d applies some damping.

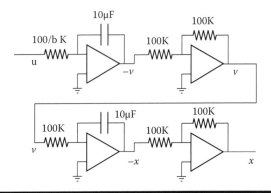

Figure 2.8   Simulating a second order system.

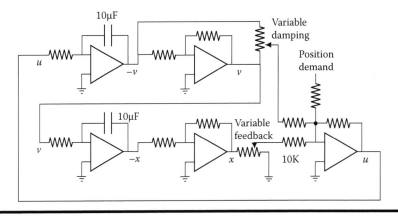

Figure 2.9   Second order system with variable feedback.

If we connect a potentiometer resistor between the outputs v and $-v$ of the operational amplifiers relating to v, then the wiper can pick off a term of either sign as shown in Figure 2.9. Thus we can add positive or negative damping. Another potentiometer connected between the supplies can represent the position demanded, while a further one feeds back the position error. The potentiometer signals are mixed in an invertor (which doubles as a summer), giving an output u which is applied to the input of the v integrator.

2.10 Closed Loop Equations

Now in both the digital and analog cases, we see that the differential state equations merely list the inputs to each integrator in terms of system inputs and the state variables themselves. If we look at the integrator inputs of our closed loop system, we see that dx/dt is still v, but now dv/dt has a mixture of inputs.

When we substitute the feedback value $f.(x_{\text{demand}} - x) - d.v$ for u, we find that the equations have become:

$$\frac{dx}{dt} = v$$

and

$$\frac{dv}{dt} = b(f.(x_{\text{demand}} - x) - d.v)$$

or in matrix terms

$$\begin{bmatrix} \dot{x} \\ \dot{v} \end{bmatrix} = \begin{bmatrix} 0 & 1 \\ -bf & -bd \end{bmatrix} \begin{bmatrix} x \\ v \end{bmatrix} + \begin{bmatrix} 0 \\ bf \end{bmatrix} x_{\text{demand}}.$$

We see that applying feedback has turned our second order system, described by a matrix equation:

$$\dot{\mathbf{x}} = \mathbf{Ax} + \mathbf{Bu}$$

into another system, describable in exactly the same form except that the **A** and **B** matrices are different, and we have a new input. We can perhaps sum up the problem of control as follows:

"By feedback, we can change the matrices describing the system. How do we achieve a set of matrices which we like better than the ones we started with?"

Chapter 3

Simulation with JOLLIES: JavaScript On-Line Learning Interactive Environment for Simulation

3.1 Introduction

We have already seen how useful simulation can be for analyzing linear systems. When the system is non-linear, and most of them are, simulation is an essential tool.

The examples in the last chapter were cobbled together so that you only had to type in a minimum quantity of code to achieve some sort of graphical output. By using the "Jollies" framework, you can concentrate on the control aspects while most of the graphics housekeeping is taken care of.

The acronym Jollies stands for "Javascript on-line learning interactive environment for simulation." Early examples were mounted on a website www.jollies. com, and I suspect that many of the visitors expected something of a different nature.

A significant feature of the technique is two text windows, as shown in Figure 3.1. When the web page is opened, this text can be changed at will. When the "run" button is clicked, the text in the first window is "evaluated," in other words, it is run as computer code. It is used to initialize variables and parameters, to set or change the step-length, and to initialize and scale the graphics window.

The code in the second window is then run repeatedly, each time advancing the simulation by time *dt* and plotting the results on a graphics applet or moving images around the screen.

By clicking "buttons" on the page, inputs and parameters can be changed during the running of the model.

We can see how various components fit together by examining a more complete simulation of the water-butt problem.

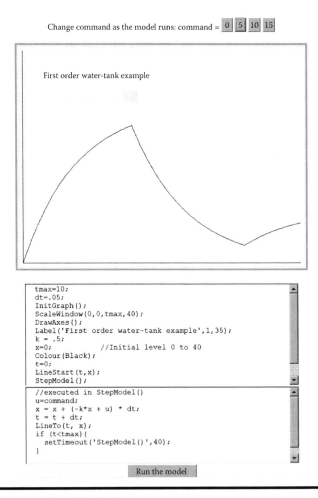

Figure 3.1 Water-butt simulation.

3.2 How a JOLLIES Simulation Is Made Up

Firstly, there is the usual heading of a web page

```
<html>
<head>
  <meta http-equiv="author" content="John Billingsley USQ">
  <title>Simulation Window</title>
```

Then we use Javascript to define many of the variables. Since any variables defined in the "initial" text window would not normally be shared by the "model" code, these variables must be defined at the head of the page.

```
<script LANGUAGE="javaScript">
var apheight=400;
var apwidth=640;
var g=0;
var dt=.04;
var k=0;
var t=0;
var x=0;
var u=15;
var command=15;
var tmax=10;
var xmin=0.0;
var xmax=1.0;
var ymin=-1;
var ymax = 1;
var xsc=60.0;
var ysc=160.0;
var Red=new Array(255,0,0);
var Blue=new Array(0,0,255);
var Green=new Array(0,255,0);
var Black=new Array(0,0,0);
var Cream=new Array(255,255,200);
var Dull=new Array(200,200,150);
var firsttime=true;
```

The applet is then initialized and its properties and methods copied to the local variable g.

The applet is filled with a cream background color, with a shaded border.

```
function InitGraph() {
  document.Graph.init();
  g=document.Graph.g;
```

```
  Colour(Dull);
  g.fillRect(0,0,apwidth,apheight);
  Colour(Cream);
  g.fillRect(5,5,apwidth-10,apheight-10);
}
```

By using a function to scale the coordinates, we can plot the variables without worrying about scales or offsets

```
function ScaleWindow(a,b,c,d) {
  xmin=a;
  ymin=b;
  xmax=c;
  ymax=d;
  xsc=(apwidth-40)/(xmax-xmin);
  ysc=(apheight-40)/(ymin-ymax);
  g.translate(20-xmin*xsc,20-ymax*ysc);
}
```

The LineStart() routine moves the pen and marks a single dot.

```
function LineStart(p,q) {
  oldx=p*xsc;
  oldy=q*ysc;
  g.drawLine(oldx,oldy,oldx+1,oldy);
}
```

LineTo() draws a line from the pen's old position. These coordinates are those of the model, not the scaled pixel coordinates of the graph.

```
function LineTo(p,q) {
  newx=p*xsc;
  newy=q*ysc;
  g.drawLine(oldx,oldy,newx,newy);
  oldx=newx;
  oldy=newy;
}
```

Box() and BoxFill() draw scaled rectangles.

```
function Box(a,b,c,d) {
  g.drawRect(a*xsc,b*ysc,c*xsc,-d*ysc);
}
```

```
function BoxFill(a,b,c,d) {
  g.fillRect(a*xsc,(b+d)*ysc,c*xsc,-d*ysc);
}
```

```
function Colour(c) {
  document.Graph.setRGB(c[0],c[1],c[2]);
}

function ColourRGB(r,g,b) {
  c=new Array(r,g,b);
  Colour(c);
}
```

The final graphics functions give an easy way to draw the axes and to place a label at a chosen position.

```
function DrawAxes() {
  Colour(Blue);
  LineStart(xmin,0);
  LineTo(xmax,0);
  LineStart(0,ymin);
  LineTo(0,ymax);
}

function Label(a,b,c) {
  g.drawString(a,b*xsc,c*ysc);
}
```

Finally, we must define a way to execute the code in the model window, in a way that can be called at fixed intervals of time by the SetTimeout() function.

```
function StepModel() {
  eval(document.panels.model.value);
}

</script>
</head>
```

Now, having set up the "head" of the page we must position the text and features that the user will see. At the top is a "form" with the buttons that will change the command.

```
<body>
<center>
<FORM name=inputs>
Change command as the model runs: command =
<INPUT TYPE="button"
  NAME="u0"
  VALUE=" 0 "
  onClick="command=0;" >
<INPUT TYPE="button"
```

```
  NAME="u5"
  VALUE=" 5 "
  onClick="command=5;" >
<INPUT TYPE="button"
  NAME="u10"
  VALUE="10"
  onClick="command=10;" >
<INPUT TYPE="button"
  NAME="u15"
  VALUE="15"
  onClick="command=15;" >
<br>
</form>
```

Then we place the applet on the page. Obviously we can define it to be any size and shape that will fit on the page, as long as we set the variables *apheight* and *apwidth* at the top to be the same as the height and width settings.

```
<applet
  code=Graph.class
  name=Graph
  width=640
  height=400>
</applet>
```

Now we have the "form" with the two text panels or TEXTAREAs. The first contains the code for setting up the display and initializing the variables.

```
<form name=panels>
<TEXTAREA NAME="initial" rows=11 cols=60>
  tmax=10;
  dt=.05;
  InitGraph();
  ScaleWindow(0,0,tmax,40);
  DrawAxes();
  Label('First order water-tank example',1,35);
  k = .5;
  x=0; //Initial level 0 to 40
  Colour(Black);
  t=0;
  LineStart(t,x);
  StepModel();

</TEXTAREA>
<BR>
```

The second contains the code for stepping the simulation and plotting the result.

```
<TEXTAREA NAME="model" rows=7 cols=60>
  //executed in StepModel()
  u=command;
  x = x + (-k*x + u) * dt;
  t = t + dt;
  LineTo(t, x);
  if (t<tmax){
   setTimeout('StepModel()',40);
  }

</TEXTAREA>
<BR>
```

Provided we make sure that any variables have been declared in the <head> code, we can edit the contents of these windows to simulate almost any system we can think of.

Now we have the button to click to start it all running.

```
<input TYPE="button"
  NAME="doit"
  VALUE="Run the Model"
  onClick="
          window.scrollTo(1,1);
          setTimeout('eval(document.panels.initial.
          value);',500);
          " >

<p>
</center>
</form>
```

The page is scrolled to the top and we wait for the start button.

```
<script>
  window.scrollTo(1,460);
</script>
</body>
</html>
```

You can copy and save this code, or find it already saved on the book's website as www.esscont.com/3/2-ButtAction.htm.

3.3 Moving Images without an Applet

An alternative to plotting a graph is to animate a simulation, by moving objects around the screen. In the example below, the object is the simple word "Bounce," but any graphic that can be loaded into a web page can be used.

The system in this case represents a ball, falling under gravity while at the same time traveling horizontally. In each interval *dt*, the vertical velocity *vy* will change by *g*dt*, while the horizontal *vx* velocity stays the same. The vertical position will change by *vy*dt* and the horizontal position by *vx*dt*.

If the vertical position falls below zero and the velocity is downwards, the ball must bounce. We have a system that is non-linear in the extreme! The vertical velocity is reversed in sign, but is also reduced to 0.8 of its magnitude. In the simulation, the horizontal velocity is also reduced in each bounce.

The complete code is:

```
<html>
<head>
   <meta http-equiv="author" content="John Billingsley USQ">
   <title>Moving Image</title>
<script LANGUAGE="javaScript">
var x;
var vx;
var y;
var vy;
var dt=.1;
var g=-10;
var running=false;

function Move(a,b,c) {
  a.style.left=b;
  a.style.top=500-c;
}

function step(){
  eval(document.panels.prog.value);
  if(running){setTimeout('step();',20);}
  else{document.panels.doit.value='Run';};
}

</script>
</head>

<body>
<div id="ball" style="position:absolute; left:0; top:0">
Bounce
</div>

<center>
<form name=panels>
<input TYPE="button"
  NAME="doit"
```

```
        VALUE="Run "
        onClick="
                if (doit.value=='Run'){
                  doit.value='Stop';
                  running=true;
```

Do the initialization here, rather than in a separate text box

```
                  x=100;
                  y=400;
                  vx=10;
                  vy=0;
                  step();
                }else{
                  doit.value='Run';
                  running=false;
                };
                " >
<br>
<TEXTAREA NAME="prog" rows=15 cols=25>
//Step

y=y+vy*dt;
vy=vy+g*dt;
x=x+vx*dt;

Move(ball,x,y);

if((y<0)&(vy<0)){
  vy=-.8*vy;
  vx=.8*vx;
};

</TEXTAREA>
</form>
</center>

<script>
  var ball=document.getElementById("ball");
</script>

</body>
</html>
```

This is saved as www.esscont.com/3/3-bounce.htm.

Instead of a text word, we can move a picture around the screen with equal ease. In the next chapter we will set up a simulation to represent a laboratory experiment.

3.4 A Generic Simulation

In any simulation, a decision must be made about naming the variables. They can be chosen and defined to represent the "intuitive" state variables, with names that clearly denote position, velocity, temperature, or such, or the states can be represented by components of a generic vector, **x**, represented in code by lines such as:

```
x[0]=x[0]+x[1]*dt
```

Now if our system is linear and can be represented by a matrix state equation

$$\dot{x} = Ax + Bu$$

we can perform the calculations in the form of matrix operations. The format of the code is always the same, while different values are given to the matrices to represent different systems. The dimensions of the matrices can be increased to represent higher orders, provided the counting limits in the loops are increased to match.

The first example in this chapter can be modified to represent a linear second order system, such as a position control, as in the following.

In the code in the <head>, additional variables are declared to represent matrices **A** and **B** and state vector **x**, with

```
var A=new Array(2);
var B=new Array(2);
var x=new Array(2);
```

The "initialize" panel code is now replaced, giving values to the matrices and initial conditions to the state:

```
tmax=10;
  dt=.01;
  InitGraph();
  ScaleWindow(0,-10,tmax,30);
  DrawAxes();
  Label('Second order position control',1,25);
  A[0]=Array( 0, 1); //A-matrix of the system
  A[1]=Array(-6,-2);
  B=Array( 0, 6);
  x=Array( 0, 0); //Initial position and velocity
  Colour(Black);
  t=0;
  LineStart(t,x[0]);
  StepModel();
```

while the code in the "model" panel is also replaced, now having a double loop for the matrix multiplication.

```
for(var i=0;i<=1;i++){
  for(var j=0;j<=1;j++){
    x[i]+=A[i][j]*x[j]*dt;
  }
  x[i]+=B[i]*u*dt;
}
t = t + dt;
LineTo(t, x[0]);
if (t<tmax){
  setTimeout('StepModel()',10);
}
```

Take care, though. The integration accuracy still depends on a small enough value for *dt*. Later we will see matrix methods for an exact simulation.

This is saved as www.esscont.com/3/4-Generic.htm.

Another version has separated the JavaScript plotting functions into a separate file, jollies.js, so that the simulation file can contain much less "housekeeping."

It is saved at www.esscont.com/3/5-Include.htm.

Q 3.4.1

Describe the response of this simulation. Is it (a) underdamped, (b) critically damped, or (c) overdamped? Does it overshoot?

Chapter 4

Practical Control Systems

4.1 Introduction

An electric iron manages to achieve temperature control with one single bimetal switch. Guiding the space shuttle requires rather more control complexity. Control systems can be vast or small, can aim at smooth stability or a switching limit cycle, can be designed for supreme performance, or can be the cheapest and most expedient way to control a throwaway consumer product. So where should the design of a controller begin?

There must first be some specification of the performance required of the controlled system. In the now familiar servomotor example, we must be told how accurately the output position must be held, what forces might disturb it, how fast and with what acceleration the output position is required to move. Considerations of reliability and lifetime must then be taken into account. Will the system be required to work only in isolation, or is it part of a more complex whole?

A simple radio-controlled model servomotor will use a small DC motor, with a potentiometer to measure output position. For small deviations from the target position the amplifier in the loop will apply a voltage to the motor proportional to error, and with luck the output will achieve the desired position without too much overshoot.

An industrial robot arm requires much more attention. The motor may still be DC, but will probably be of high performance at no small cost. To ensure a well-damped response, the motor may well have a built-in tachometer that gives a measure of its speed. A potentiometer is hardly good enough, in terms of accuracy or lifespan; an incremental optical transducer is much more likely—although some systems have both. Now it is unlikely that the control loop will be closed merely by a simple amplifier; a computer is almost certain to get into the act. Once this level of complexity is reached, position control examples show many common features.

When it comes to the computer that applies the control, it is the control strategy that counts, rather than the size of the system. A radio-telescope in the South of England used to be controlled by two mainframes, with dubious success. They were replaced by two personal microcomputers, with a purchase cost that was no more than a twentieth of the mainframes' annual maintenance cost, and the performance was much improved.

4.2 The Nature of Sensors

Without delving into the physics or electronics, there are characteristics of sensors and actuators that are fundamental to many of the control decisions. Let us try to put them into some sort of order of complexity.

By orders of magnitude, the most common sensor is the thermostat. Not only it is the sensor that detects the status of the temperature, it is also the controller that connects or disconnects power to a heating element. In an electric kettle, the operation is a once-and-for-all disconnection when the kettle boils. In a hot-water urn, the connection re-closes to maintain the desired temperature.

Even something as simple as this needs some thought. To avoid early burnout of the contacts, the speed of the on–off switching cycle must be relatively slow, something usually implemented by hysteresis in the sensor. If the heater is a room heater, should the thermostat respond just to the temperature of the room, or should the temperature of the heater itself play a part? In the latter case, the limit cycle of the room temperature itself will be much reduced. On the other hand, the outside temperature will then have much more influence on the mean temperature.

Many other sensors are also of a "single bit" nature. Limit switches, whether in the form of microswitches or contactless devices, can either bring a movement to an end or can inhibit movement past a virtual end stop. Similar sensors are at the heart of power-assisted steering. When the wheel is turned to one side of a small "backlash" in the wheel-to-steering linkage, the "assistance" drives the steering to follow the demand, turning itself off when the steering matches the wheel's demand.

When we wish to measure a position we can use an "incremental encoder." The signals are again based on simple on–off values, but the transitions are now counted in the controller to give a broad range of values. If the transducer senses equal stripes at one millimeter intervals, the length is theoretically limitless but the position cannot be known to better than within each half millimeter stripe.

Simple counting is adequate if the motion is known to be in a single direction, but if the motion can reverse then a "two phase" sensor is needed. A second sensor is mounted quarter of a cycle from the first, quarter of a millimeter in this case, so that the output is obtained as pairs of levels or bits. The signals are shown in Figure 4.1. Now the sequence 00 01 11 10 will represent movement in one direction, while 00 10 11 01 will represent the opposite direction. The stripes can be mounted on a rotating motor, just as easily as on a linear track.

This type of sensor has the advantage of extreme simplicity and is at the heart of the "rolling ball" computer mouse. It has the disadvantage that the sensing is of relative motion and there is no knowledge of the position at switch-on. Since the increments are quantized, in this example to a precision of a quarter of a millimeter, it is impossible to control to a tighter resolution.

It is possible to measure an "instant position" by sensing many stripes in parallel. With 10 sensors, the rotation angle of a disk can be measured to one part in a 1000. But the alignment of the stripes must be of better accuracy than one part in a 1000, resulting in a very expensive transducer.

The classical way to signal the value of a continuous measurement is by means of a varying voltage or current. A common position transducer is the potentiometer. A voltage is applied across a resistive track and a moving "wiper" picks off a voltage that is proportional to the movement along the track. This is shown in Figure 4.2. There is no quantization of the voltage, but in all except the most expensive of potentiometers there will be "gritty" noise as the wiper moves. There is also likely to be non-linearity in the relationship between voltage and position.

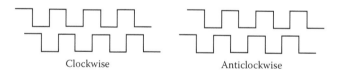

Clockwise   Anticlockwise

Figure 4.1   Two-phase encoder waveforms.

Figure 4.2   A potentiometer can measure position.

Non-contact variations on the potentiometer principle include Hall-effect devices that sense the varying angle of a magnetic field and transformer-based devices such as the "E and I pickoff" or the LVDT "Linear Variable Differential Transformer" in which movement changes the coupling between an alternating field and a detection coil.

To measure force, a strain-gauge relies on the variation of the resistance of a thin metallic film as it is stretched. It is just one of the many devices in which the quantity is measured by means of displacement of one sort or another, in this case by the bend in a bracket.

Even though all these measurements might seem to be continuous and without any steps, quantization will still be present if a digital controller is involved. The signal must be converted into a numerical value. The analog-to-digital converter might be "8 bit," meaning that there are 256 steps in the value, up to 16-bit with 65,536 different levels. Although more bits are possible beyond this, noise is likely to limit their value.

4.3 Velocity and Acceleration

It is possible to measure or "guess" the velocity from a position measurement. The crude way is to take the difference between two positions and divide by the time between the measurements. A more sophisticated way is by means of the high-pass filter that we will meet later in the book.

Other transducers can give a more direct measurement. When a motor spins, it generates a "back emf," a voltage that is proportional to the rotational velocity. So the addition of a small motor to the drive motor shaft can give a direct measurement of speed. This sensor is commonly known as a tachometer or "tacho."

The rotation of a moving body is measured by a *rate-gyro*. In the past, this took the form of a spinning rotor that acted as a gyroscope. A rotation, even a slow one, would cause the gyroscope to precess and twist about a perpendicular axis. This twist was measured by an "E and I pickoff" variable transformer. Today, however, the "gyro" name will just be a matter of tradition. In a tiny vibrating tuning fork, rotation about the axis causes the tines to exhibit sideways vibrations. These can be picked up to generate an output.

An acceleration or tilt can be made to produce a displacement by causing a mass to compress or stretch a spring. Once more a displacement transducer of one sort or another will produce the output.

4.4 Output Transducers

When the output is to be a movement, there is an almost unbelievable choice of motors and ways to drive them.

In the control laboratory, the most likely motor that we will find is the permanent-magnet DC motor. These come in all sizes and are easily recognized in their automotive applications, such as windscreen wipers, window winders, and rear-view mirror adjusters. Although a variable voltage could be applied to drive such a motor at a variable speed, the controller is more likely to apply a "mark-space" drive. Instead of switching on continuously, the drive switches rapidly on and off so that power is applied for a proportion of the time.

To avoid the need for both positive and negative power supplies, a circuit can be used that is called an "H-bridge." The principle is shown in Figure 4.3 and a suitable circuit can be found on the book's website. With the motor located in the cross-bar of the H, either side can be switched to the single positive supply or to ground, so that the drive can be reversed or turned off. A and D on with B and C off will drive the motor one way while B and C on with A and D off will drive it the other way. By switching B and D on while A and C are off, braking can be applied to stop the motor.

Another hobbyist favorite is the "stepper motor." There are four windings that can be thought of as North, South, East, and West. Each has one end connected to the supply, so that it can be energized by pulling the other pin to ground. Just like a compass, when the North winding is energized, the rotor will move to a North position. The sequence N, NE, E, SE, S, SW, W, NW, and back to North will move the rotor along to the next North position. Yes, it is the "next" North because there are many poles in the motor and it will typically take 50 of these cycles to move the rotor through one revolution.

Stepper motors are simple in concept, but they are lacking in top speed, acceleration, and efficiency. They were useful for changing tracks on floppy disks, when these were still floppy. Large and expensive versions are still used in some machine tools, but a smaller servomotor can easily outperform them.

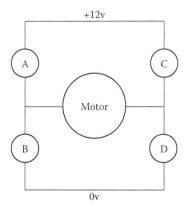

Figure 4.3　Schematic of H-bridge.

A DC motor typically uses brushes for commutation, for changing the energization of the windings so that the rotor is continually urged onwards. "Brushless" motors use electronics to do this, to gain extended life for applications as simple as cooling fans for personal computers.

Electric motors are far from being the only actuators. Hydraulic and pneumatic systems allow large forces to be controlled by switching simple solenoid valves.

In general, the computer can exercise its control through the switching of a few output bits. It is seldom necessary to have very fine resolution for the output, since high gains in the controller mean that a substantial change in the output level corresponds to a very small change in the error at the input.

4.5 A Control Experiment

A practical control task is an essential part of any undergraduate course in control. There is a vast range of choice for the design of such a system. Something must move, and an important decision must be made on the measurement of that movement. Whether linear or rotary, the measurements can be split into relative and absolute, into continuous or discrete.

An inverted pendulum experiment is remarkably easy to construct and by no means as difficult to control as the vendors of laboratory experiments would have you believe. Some construction details can be found on this book's website. Follow the link via www.esscont.com/4/pendulum.htm. Such a system has the added advantage that the pendulum can be removed to leave a practical position control system.

With such a system, the features of position control can be explored that are so often hidden in laboratory experiments. Of course settling time and final accuracy are important, but for the design of an industrial controller it is necessary to test its ability to withstand a disturbing force. All too often, the purchased experiments protect the output with a sheet of perspex, so that it is impossible for the experimenter to test the "stiffness" of he system.

The next best thing to a practical experiment is a simulation. To be useful it must obey the same laws and constraints, must allow us to experiment with a great variety of control algorithms and have a way to visualize the results.

In preparation for coming chapters, let us set up a simulation on which we can try out some pragmatic strategies.

In Chapter 2, a simple position controller was introduced, but with few practical limitations. Let us look at how an experiment might be constructed. A motor with a pulley and belt accelerates a load. The position could be measured with a multi-turn potentiometer (Figure 4.4).

Our input controls the acceleration of the motor. We can define state variables to be the position and velocity, x and v.

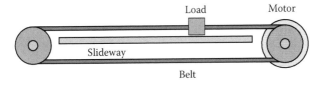

Figure 4.4   Position controller.

The first state equation is very simple. It simply states that rate-of-change of position is equal to the velocity!

$$\frac{dx}{dt} = v.$$

The second one expresses the acceleration in terms of the input and the velocity,

$$\frac{dv}{dt} = bu - av$$

since a typical motor has inbuilt damping. For now we will ignore this and set a to zero.

So now we need to set up a believable simulation that will show us the result of any feedback that we may apply. The feedback will involve making the input depend on the position and velocity, or maybe just on the position alone if we have no velocity signal available.

Load the software from www.esscont.com/4/position.htm. On the screen you will see a block moved along a "belt" by the code:

```
u= -2*x-5v;
v = v + u*dt;
x = x + v*dt;
```

and you will see that it is rather slow to settle.

Change the first line to

```
u= -20*x-5*v;
```

and it settles much more quickly.

Now try

```
u= -100*x-20*v;
```

and the performance is even better.

But this model is lacking many important features that we will add in the chapters to come. Eventually in Chapter 9, it will enable us to model the inverted pendulum, but there are many lessons in non-linear theory to be learned first.

Chapter 5

Adding Control

5.1 Introduction

Some approaches to control theory draw a magical boundary between open loop and closed loop systems. Yet, toward the end of Chapter 2 we saw that a similar-looking set of state equations described either open or closed loop behavior. Our problem is to find how to modify these equations by means of feedback, so that the system in its new form behaves in a way that is more desirable than it was before.

We will see how to simulate a position control system with the ability to inject real-time disturbances. We will meet mathematical methods for deciding on feedback values, though the theory is often misapplied. Because of the close relation between state equations and with simulation, however, we can use the methods of this chapter to set up simulations on which we can try out any scheme that comes to mind.

5.2 Vector State Equations

It is tempting to assert that every dynamic system can be represented by a set of state equations in the form:

$$\dot{\mathbf{x}} = \mathbf{Ax} + \mathbf{Bu} \tag{5.1}$$

where **x** and **u** are vectors and where there are as many separate equations as **x** has components.

Unfortunately, there are many exceptions. A sharp switching action cannot reasonably be expressed by differential equations of any sort. A highly non-linear system will only approximate to the above form of equations when its disturbance is very small. A pure time delay, such as the water traveling through the hose of your bathroom shower, will have a variable for the temperature of each drop of water that is in transit. Nevertheless the majority of systems with which the control engineer is concerned will fall closely enough to the matrix form that it becomes a very useful tool indeed.

For all its virtues in describing the way in which the state of the system changes with time, Equation 5.1 is only half of the story. Suppose that we take a closer look at the motor position controller shown in Figure 4.4, with a potentiometer to measure the output position and some other constants defined for good measure.

Once again we have state variables x and v, representing the position and velocity of the output.

The "realistic" motor has a top speed; it does not accelerate indefinitely. Let us define the input, u, in terms of the proportion of full drive that is applied. If $u = 1$, the drive amplifier applies the full supply voltage to the motor.

Now when we apply full drive, the acceleration decreases as the speed increases. Let us say that the acceleration is reduced by an amount av. If the acceleration from rest is b for a maximum value 1 of input, then we see that the velocity is described by the first order differential equation:

$$\dot{v} = -av + bu. \tag{5.2}$$

The position is still a simple integral of v,

$$\dot{x} = v.$$

If we are comfortable with matrix notation, we can combine these two equations into what appears to be a single equation in a vector that has components x and v:

$$\begin{bmatrix} \dot{x} \\ \dot{v} \end{bmatrix} = \begin{bmatrix} 0 & 1 \\ 0 & -a \end{bmatrix} \begin{bmatrix} x \\ v \end{bmatrix} + \begin{bmatrix} 0 \\ b \end{bmatrix} u$$

or if we represent the vector as **x** and the input as **u** (this will allow us to have more than one component of input) the equation appears as:

$$\dot{\mathbf{x}} = \begin{bmatrix} 0 & 1 \\ 0 & -a \end{bmatrix} \mathbf{x} + \begin{bmatrix} 0 \\ b \end{bmatrix} \mathbf{u}. \tag{5.3}$$

This is an example of our typical form

$$\dot{x} = Ax + Bu.$$

Q 5.2.1

What is the top speed of this motor?

Q 5.2.2

What is the time-constant of the speed's response to a change in input?

Now for feedback purposes, what concerns us is the output voltage of the potentiometer, y. In this case, we can write $y = cx$, but to be more general we should regard this as a special case of a matrix equation:

$$y = Cx. \tag{5.4}$$

In the present case, we can only measure the position. We may be able to guess at the velocity, but without adding extra filtering circuitry we cannot feed it back. If on the other hand we had a tacho to measure the velocity directly, then the output y would become a vector with two components, one proportional to position, and the other proportional to the velocity.

For that matter, we could add two tachometers to obtain three output signals, and add a few more potentiometers into the bargain. They would be of little help in controlling this particular system, but the point is that the number of outputs is simply the number of sensors. This number might be none (not much hope for control there!), or any number that might perhaps be more than the number of state variables. Futile as it might appear at first glance, adding extra sensors has a useful purpose when making a system such as an autopilot, where the control system must be able to survive the loss of one or more signals.

The System can be portrayed as a block diagram, as in Figure 5.1.

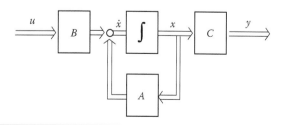

Figure 5.1 State equations in block diagram form.

5.3 Feedback

The input to our system is at present the vector **u**, which here has only one component. To apply feedback, we must mix proportions of our output signals with a command input **w** to construct the input **u** that we apply to the system.

To understand how the "command input" is different from the input **u**, think of the cruise control of a car. The input to the engine is the accelerator. This will cause the speed to increase or decrease, but without feedback to control the accelerator the speed settles to no particular value.

The command input is the speed setting of the cruise control. The accelerator is now automatically varied according to the error between the setting and actual speed, so that the response to a change in speed setting is a stable change to the new set speed.

Now by mixing states and command inputs to make the new input, we will have made

$$\mathbf{u} = \mathbf{Fy} + \mathbf{Gw}.$$

When we substitute this intothe state Equation 5.1, we obtain:

$$\dot{\mathbf{x}} = \mathbf{Ax} + \mathbf{B}(\mathbf{Fy} + \mathbf{Gw}).$$

But we also know that

$$\mathbf{y} = \mathbf{Cx}$$

so

$$\dot{\mathbf{x}} = \mathbf{Ax} + \mathbf{B}(\mathbf{FCx} + \mathbf{Gw})$$

i.e.,

$$\dot{\mathbf{x}} = (\mathbf{A} + \mathbf{BFC})\mathbf{x} + \mathbf{BGw}. \tag{5.5}$$

So the system with feedback as shown in Figure 5.2 has been reduced to a new set of equations in which the matrix **A** has been replaced by (**A** + **BFC**) and where a new matrix **BG** has replaced the input matrix **B**.

As we saw at the end of Chapter 2, our task in designing the feedback is to choose the coefficients of the feedback matrix **F** to make (**A** + **BFC**) represent something with the response we are looking for.

If **B** and **C**, the input and output matrices, both had four useful elements then we could choose the four components of **F** to achieve any closed loop state matrix

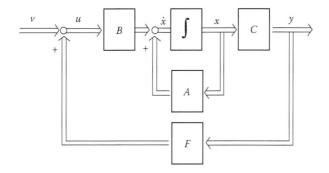

Figure 5.2 Feedback in block diagram form.

we wished. Unfortunately in this case, with a single input and a single output that is proportional to position, they have degenerated to:

$$\mathbf{B} = \begin{bmatrix} 0 \\ b \end{bmatrix}$$

and

$$\mathbf{C} = \begin{bmatrix} c & 0 \end{bmatrix}$$

so that **F** has reduced to a single scalar *f*. We see that:

$$\mathbf{BFC} = \begin{bmatrix} 0 & 0 \\ bfc & 0 \end{bmatrix}$$

so that

$$\mathbf{A} + \mathbf{BFC} = \begin{bmatrix} 0 & 1 \\ bfc & -a \end{bmatrix}.$$

To make it clear what this means, let us look at the problem using familiar techniques.

5.4 Another Approach

We can attack the position control example in the "traditional" way as follows. We found a differential Equation 5.2 for the motor speed

$$\dot{v} = -av + bu.$$

We can use the relationship between velocity v and position x to see it as

$$\ddot{x} = -a\dot{x} + bu. \tag{5.6}$$

Now the potentiometer still gives an output y proportional to x, so

$$y = cx.$$

Feedback means mixing this potentiometer signal with another input w. We apply the result to the input so that

$$u = fy + gw.$$

Substituting this into Equation 5.6, we get:

$$\ddot{x} = -a\dot{x} + b(fcx + gv)$$

or

$$\ddot{x} + a\dot{x} - bfcx = bgw. \tag{5.7}$$

It is in the classical form of a second order differential equation.

Let us put some numbers to the constants. Suppose that the motor has a time constant of a fifth of a second, so that $a = 5$. If the initial full drive acceleration is six units per second per second, then $b = 6$. Finally, the position transducer gives us $c = 1$ volt per unit. Suppose in addition that we have chosen to set the feedback coefficient $f = -1$ and the command gain $g = 1$. Then Equation 5.7 becomes

$$\ddot{x} + 5\dot{x} + 6x = 6w. \tag{5.8}$$

If w is constant, this has a "Particular Integral" $x = w$. To complete the general solution, we need to find the "Complementary Function," the solution of

$$\ddot{x} + 5\dot{x} + 6x = 0. \tag{5.9}$$

Suppose that the solution is of the form:

$$x = e^{mt}.$$

Then

$$\dot{x} = me^{mt}$$

and

$$\ddot{x} = m^2 e^{mt}.$$

When we substitute these into Equation 5.9, we get:

$$m^2 e^{mt} + 5me^{mt} + 6e^{mt} = 0$$

and we can take out the exponential as a factor to get

$$(m^2 + 5m + 6)e^{mt} = 0.$$

Since the exponential will not be zero unless *mt* has a value of minus infinity, we have:

$$m^2 + 5m + 6 = 0$$

and we can solve for the value of *m*, to find

$$m = -2 \quad \text{or} \quad m = -3.$$

The Complementary Function is a mixture of two exponentials

$$x = Ae^{-2t} + Be^{-3t}$$

where *A* and *B* are constants, chosen to make the function fit the two initial conditions of position and velocity.

5.5 A Change of Variables

Suppose that we define two new variables:

$$w_1 = \dot{x} + 2x$$

$$w_2 = \dot{x} + 3x. \tag{5.10}$$

The reason becomes obvious when we realize that

$$\dot{w}_1 + 3w_1 = \ddot{x} + 5\dot{x} + 6x$$

and also

$$\dot{w}_2 + 2w_2 = \ddot{x} + 5\dot{x} + 6x.$$

So our second order differential equations have been turned into a pair of first order equations in which the two variables are only connected by the right-hand-sides of the equations:

$$\dot{w}_1 + 3w_1 = 6w$$

$$\dot{w}_2 + 2w_2 = 6w$$

We can express these equations in the form of a single matrix equation:

$$\begin{bmatrix} \dot{w}_1 \\ \dot{w}_2 \end{bmatrix} = \begin{bmatrix} -3 & 0 \\ 0 & -2 \end{bmatrix} \begin{bmatrix} w_1 \\ w_2 \end{bmatrix} + \begin{bmatrix} 6 \\ 6 \end{bmatrix} w \tag{5.11}$$

and clearly there is something special about the diagonal form of the system matrix.

We can make the matrices work even harder for us by expressing the new state variables in the form of a transformation of the original variables x and v.

Equations 5.10 can be rewritten as:

$$w_1 = 2x + v$$

$$w_2 = 3x + v$$

or

$$\begin{bmatrix} w_1 \\ w_2 \end{bmatrix} = \begin{bmatrix} 2 & 1 \\ 3 & 1 \end{bmatrix} \begin{bmatrix} x \\ v \end{bmatrix}. \tag{5.12}$$

This transformation $\mathbf{w} = \mathbf{Tx}$ has an inverse $\mathbf{x} = \mathbf{T}^{-1}\mathbf{w}$ that expresses $\mathbf{x}$ in terms of $\mathbf{w}$.

$$\begin{bmatrix} x \\ v \end{bmatrix} = \begin{bmatrix} -1 & 1 \\ 3 & -2 \end{bmatrix} \begin{bmatrix} w_1 \\ w_2 \end{bmatrix}. \tag{5.13}$$

In general, if the set of matrix state equation in vector $\mathbf{x}$ is given by

$$\dot{\mathbf{x}} = \mathbf{Ax} + \mathbf{Bu}$$

and we have new variables which are the components of **w**, where we know the transformation that links **w** and **x**, we have:

$$\dot{\mathbf{w}} = T\dot{\mathbf{x}}$$

so

$$\dot{\mathbf{w}} = TA\mathbf{x} + TB\mathbf{u}$$

i.e.,

$$\dot{\mathbf{w}} = TAT^{-1}\mathbf{w} + TB\mathbf{u}. \tag{5.14}$$

Given any one set of state variables, we can produce a new set of state equations in terms of transformation of them. The variety of equations is literally infinite for any one problem. However, just one or two forms will be of particular interest, as we will see later.

Q 5.5.1

Use the transformation and its inverse, given by Equations 5.12 and 5.13. Starting from Equation 5.15 and substituting values in Equation 5.14 show that we arrive at Equation 5.11.

$$\begin{bmatrix} \dot{x} \\ \dot{v} \end{bmatrix} = \begin{bmatrix} 0 & 1 \\ -6 & -5 \end{bmatrix} \begin{bmatrix} x \\ v \end{bmatrix} + \begin{bmatrix} 0 \\ 6 \end{bmatrix} w \tag{5.15}$$

where w (note the lower case, not a vector) represents the demanded position.

This gives some insight into the way to analyze the response of a linear system, but is of little use in designing a controller.

When we come to the task of designing a real controller for a real motor, the fact that there is a limit on the drive that can be applied has a major effect on the choice of feedback. We will have to find ways of analyzing systems that are not linear.

5.6 Systems with Time Delay and the PID Controller

When there is a *transport delay*, such as in the hose of a bathroom shower, our neat system of state equations breaks down. When we adjust the mixer handle, there is no change or rate-of-change or any higher derivative of the output temperature until we have waited a second or two.

Not only does this present us with a difficult simulation problem, it gives a tough control problem too. If we want to use a continuous controller, we have to consider *integral control*.

Next time you stand in the shower, apply proportional control with a high gain. If the water is too hot, turn the mixer handle to "cold." After a second or two the cold water will hit you and you turn the handle to "hot." Beware of getting scalded! The system is unstable.

The only safe strategy is to use a very low gain, only moving the handle slightly away from the nominal position. But that might leave you with the water rather too cool, even though the handle is set a little above the mid position. So you apply *integral action*. As you start to shiver, you move the handle slowly toward "hot" until the right temperature is found. The handle must move slowly, because you have to wait for the effect of any change to reach the sensor, your skin.

To demonstrate this sort of problem, many years ago I devised an experiment for the control laboratory of Cambridge University Engineering Department. It is shown in Figure 5.3.

Water flows over a heating element and via a pipe to the output, where its temperature is measured. An input knob sets the target temperature, and power is supplied to the heater in proportion to the control action. In the analysis, temperatures are measured relative to the ambient value. The open loop response to a step change of input is shown in Figure 5.4.

Suppose, that we apply just proportional control, so that the heater drive is proportional to the temperature error.

If the heater gain is such that in the steady state

$$T_{out} = Au$$

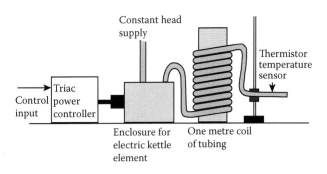

Figure 5.3 Water heater experiment.

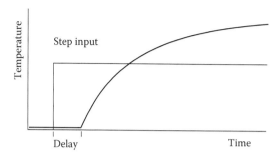

Figure 5.4 Temperature response to a step change of input.

and we apply

$$u = k(T_{\text{demanded}} - T_{\text{out}})$$

then in the steady state (if there is one) algebra tells us that:

$$T_{\text{out}} = \frac{kA}{1 + kA} T_{\text{demanded}}.$$

For any setting above ambient, there must always be a non-zero error to ensure that the heater supplies power. Unless kA is large, the temperature will not accurately follow the demand.

But it is found that for any but a modest value of kA, the system becomes unstable. Some additional method must be used to reduce the error, if accurate control is needed. Since the flow can vary, which in turn will affect the value of A, recalibrating the demand knob will not be enough!

The accepted solution is to use integral control. In addition to the proportional feedback term, an integrator is driven by the error signal and its output is added to the drive, changing slowly enough that stability is not lost. The eventual error will now be reduced to zero, since the integrator can then "take the strain" of maintaining a drive signal. The method has its drawbacks. The integrator has increased the order of the system, which does not help the problem of ensuring stability. There is also the problem of *integral wind-up*.

Suppose that after the system has settled, the demand is suddenly increased as shown in Figure 5.5. All the time that the system warms to catch up with the demand, the integrator value will be driven upward by the error. By the time the system reaches the new target temperature, the integrator will have reached a value well in excess of that needed to sustain it. The temperature will then overshoot, so that the error can wind the integrator back down again. The error–time curve must have an area of excess temperature nearly equal to the area of the error when warming.

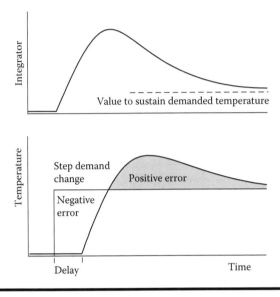

Figure 5.5 Time response showing *integral windup*.

Integral wind-up can be held in check by ensuring that the output of the integrator is limited at a level only slightly in excess of the greatest value needed. A limiter can also be applied at the input of the integrator, to reduce the wind-up effect of a large error.

In a classical PID controller, there is a single measured error signal. This is "differentiated" by some sort of high pass filter and the resulting D term is added to the P term that is proportional to the error. The error is also integrated to give the I term, lending the middle letter to the acronym PID. There is a great art to *tuning* a PID controller.

In a digital controller, more sophisticated algorithms can ensure that integration does not start until the rate-of-change of output has fallen below a certain level. The algorithm can be tailored to the physical needs of the system. You can experiment with the simulation that we are about to construct.

5.7 Simulating the Water Heater Experiment

Since we are constructing a simulation, we must define some state variables. But the whole point of this example is that it defeats the efforts of differential state equations.

The temperature in the pipe acts as a *bucket brigade*. At each computation step, the temperature of each "slice" of the pipe is passed on to the next. Differential equations are not involved.

So how many slices of pipe do we need to consider?

The tank in which the heater is mounted does have a state equation. It acts as a low-pass filter and its temperature T_{tnk} has a differential equation involving its volume V, the flow rate F, and the heater power Hu. The heat it contains is proportional to the volume times the temperature, while heat flows out at temperature times flow rate. Heat comes in via the heater.

$$\frac{d}{dt}(VT_{tnk}) = -FT_{tnk} + Hu.$$

This gives us a line of code

```
tank=tank+(-a*tank+b*u)*dt
```

where a has been preset to F/V and b is H/V.

In slicing the pipe, we need to consider the time constant of the tank and the delay of the pipe. If the tank has a time constant of 10 seconds, then a simulation step of 0.1 seconds might be appropriate. If the pipe delay is also 10 seconds, then a 100 "slices" will carry the water from slice to slice at each simulation step.

We could set up an array and shift all the contents at each time step. It is more efficient to set up the array as a "barrel." We put each new value into *pipe*[*i*] where *i* increasing to a maximum of 99, then starts again at zero. But just before doing this we extract the value that was put into that same location 100 steps earlier.

After "topping and tailing," the heart of the simulation becomes:

```
tnk=tnk+(-a*tnk+b*u)*dt;
tout=pipe[i];
pipe[i]=tnk;
i=i+1;
if(i>99){i=0;}
```

then the task remains of calculating a value for *u* and for plotting the output. For simple proportional control we would have

```
u=k*(demand-tout);
```

but we can add an integral term with

```
integral=integral+(demand-tout);
u=k*(demand-tout)+k2*integral;
```

There is a simulation to experiment with at www.esscont.com/5/heater.htm.

Chapter 6

Systems with Real Components and Saturating Signals—Use of the Phase Plane

6.1 An Early Glimpse of Pole Assignment

At the end of Chapter 4, we introduced a "moving picture" simulation. We used a motor with acceleration as the input and with both position and velocity feedback.

Load the software from www.esscont.com/4/position.htm again. The original code for stepping the model was

```
u= -2*x-5*v;
v = v + u*dt;
x = x + v*dt;
```

which gave a slow response. We saw that changing the first line to

```
u= -20*x;
```

made a great improvement. Increasing the position feedback further without increasing the damping starts to result in overshoots, but it appears that by adding extra damping we can speed up the response indefinitely. The only limitation is imposed by the time step *dt* of our simulation, something that we can also reduce.

When we apply feedback with position gain f and velocity damping d, we set

```
u= -f*x -d*v
```

resulting in a differential equation

$$\ddot{x} + d\dot{x} + fx = 0.$$

The corresponding equation for the "roots" will be

$$m^2 + dm + f = 0.$$

But how do we go about choosing values for f and d?

We might be convinced that we want to make these roots have negative values, p and q. The equation would then be:

$$(m - p)(m - q) = 0$$

or

$$m^2(p + q)m + pq = 0.$$

So to achieve these roots, we only have to make

$$d = -(p + q)$$

and

$$f = pq.$$

We have *assigned* the poles to values p and q by working backward from the solution to construct the equation to match it.

Suppose we select values of 10 for both p and q, implying a pair of closed loop time-constants of one tenth of a second. This results in values of 100 for f and 20 for d.

Q 6.1.1

Try these values in the simulation.

It looks as though we can make the response as fast as we like.

But now we must face reality.

6.2 The Effect of Saturation

Linear theory assumes that equations hold true over any range of values so that if the position error and velocity of the position control system are doubled, then the drive will also double. Double them again and again, and the drive can be increased without limit. Common sense tells us that this is an oversimplified view of the real world; sooner or later the drive signal will reach a maximum value beyond which it cannot go.

When the motor was specified in our simulation, it was stated that the maximum drive value was 1. We must build this into the simulation by adding two lines, making the "code in the box."

```
u= -100*x -20*d;

if(u>1){u=1;}
if(u<-1){u=-1;}

v = v + u*dt;
x = x + v*dt;
```

Q 6.2.1

Make the changes and click on "run." What has gone wrong?

You will find this code when you load www.esscont.com/6/position2.htm. It is necessary to more-than-double the damping term to over 50 to avoid an overshoot.

Design of the "best" feedback can depend more on the nature of the limits of the drive than on the placement of poles. Later we will see that a requirement to withstand a disturbing force can require a position gain far in excess of 100, which will in turn exacerbate the effect of drive limitation.

6.3 Meet the Phase Plane

To deal with second order non-linear problems such as this, there is a useful graphical technique, the phase plane. Phase plane and state space are in fact one and the same; the only difference here is that we are concerned with plotting "trajectories" of the state in the form of a graph of velocity versus position. Firstly, let us explore this technique in the linear case, before the motor starts to limit.

We will continue with the now familiar example, a servo system defined by the second order differential equation

$$\ddot{x} + 5\dot{x} + 6x = 6w,$$

where w is an input value of position demand. To assess the performance of the control system, it is sufficient to set the demanded position w to zero, and to see how the system responds to a disturbance, i.e., to some initial value of position and velocity.

Firstly, let us simulate the "unlimited" system with a conventional plot of *x* against time. Have a look at www.esscont.com/6/PosPlot.htm, where

```
u = -6*(x-demand) -5*v;
```

There is the familiar overdamped response, settling without overshoot (Figure 6.1).

Now, add two lines to change the code to

```
u=-6*(x-demand) -5*v;
if (u>1){u=1;}
if (u<-1){u=-1;}
```

and you will see a slow response with huge overshoots as in Figure 6.2. A comparison of the plots tells us little about the cause. The modified code, with a caption to match, is found at www.esscont.com/6/PosPlotLim.htm.

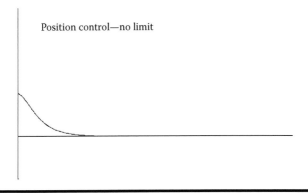

Position control—no limit

Figure 6.1   Position control without limit.

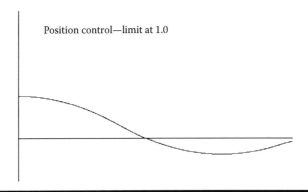

Position control—limit at 1.0

Figure 6.2   Position response with limit.

For the phase plane, instead of plotting position against time we plot velocity against position.

An extra change has been made in www.esscont.com/6/PhaseLim.htm. Whenever the drive reaches its limit, the color of the plot is changed, though this is hard to see in the black-and-white image of Figure 6.3.

For an explanation of the new plot, we will explore how to construct it without the aid of a computer. Firstly, let us look at the system without its drive limit. To allow the computer to give us a preview, we can "remark out" the two lines that impose the limit. By putting // in front of a line of code, it becomes a comment and is not executed. You can do this in the "step model" window.

Without the limit there is no overshoot and the velocity runs off the screen, as in Figure 6.4.

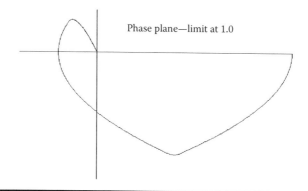

Figure 6.3   Phase plane with limit.

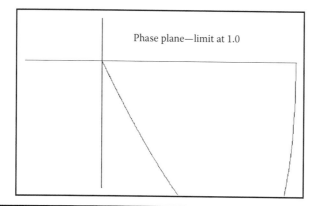

Figure 6.4   Phase-plane response without a limit.

To construct the plot by hand, we can rearrange the differential equation to set the acceleration on the left equal to the feedback terms on the right:

$$\ddot{x} = -5\dot{x} - 6x. \tag{6.1}$$

If we are to trace the track of the (position, velocity) coordinates around the phase plane, it would be helpful to know in which direction they might move. Of particular interest is the slope of their curve at any point,

$$\frac{d\dot{x}}{dx}.$$

We wish to look at the derivative of the velocity with respect to x, not with respect to time, as we usually do. These derivatives are closely related, however, since for the general function f,

$$\frac{df}{dx} = \frac{dt}{dx}\frac{df}{dt} = \frac{1}{\dot{x}}\frac{df}{dt}.$$

So, we have:

$$\text{Slope} = \frac{d\dot{x}}{dx} = \frac{1}{\dot{x}}\ddot{x}. \tag{6.2}$$

But Equation 6.1 gives the acceleration as

$$\ddot{x} = -5\dot{x} - 6x$$

so, we have

$$\text{Slope} = \frac{1}{\dot{x}}(-5\dot{x} - 6x) = -5 - 6\frac{x}{\dot{x}}. \tag{6.3}$$

The first thing we notice is that for all points where position and velocity are in the same proportion, the slope is the same. The lines of constant ratio all pass through the origin.

On the line $x = 0$, we have slope −5.
On the line $x = \dot{x}$, we have slope −11.
On the line $x = -\dot{x}$, we have slope +1, and so on.

We can make up a spider's web of lines, with small dashes showing the directions in which the "trajectories" will cross them. These lines on which the slope is the same are termed "isoclines." (Figure 6.5).

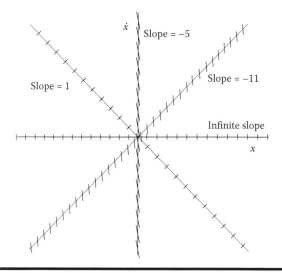

Figure 6.5   Isoclines.

An interesting isocline is the line $6x + 5\dot{x} = 0$. On this line the acceleration is zero, and so the isocline represents points of zero slope; and the trajectories cross the line horizontally.

In this particular example, there are two isoclines that are worth even closer scrutiny. Consider the line $\dot{x} + 2x = 0$. Here the slope is –2—exactly the same as the slope of the line itself. Once the trajectory encounters this line, it will lock on and never leave it. The same is true for the line $\dot{x} + 3x = 0$, where the trajectory slope is found to be –3.

Q 6.3.1

Is it a coincidence that the "special" state variables found in Section 5.5 could be expressed as $\dot{x} + 2x = 0$ and $\dot{x} + 3x = 0$, respectively?

Having mapped out the isoclines, we can steer our trajectory around the plane, following the local slope. From a variety of starting points, we can map out sets of trajectories. This is shown in Figure 6.6.

The phase plane "portrait" gives a good insight into the system's behavior, without having to make any attempt to solve its equations. We see that for any starting point, the trajectory homes in on one of the special isoclines and settles without any oscillatory behavior.

Q 6.3.2

As an exercise, sketch the phase plane for the system

$$\ddot{x} + \dot{x} + 6x = 0,$$

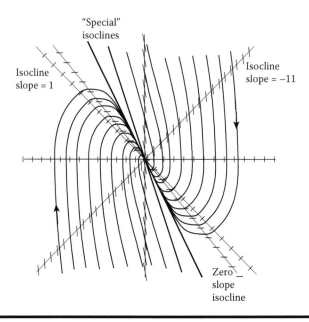

Figure 6.6   Example of phase plane with isoclines.

i.e., for the same position control system, but with much reduced velocity feedback. You will find an equation similar to Equation 6.3 for the slopes on the isoclines, but will not find any "special" isoclines. The trajectories will match the "spider's web" image better than before, spiraling into the origin to represent system responses which now are lightly damped sine-waves. Scale the plot over a range of −1 to 1 in both position and velocity.

Q 6.3.3

This is more an answer than a question. Go to the book's website. Run the simulation www.esscont.com/6/q6-3-2.htm, and compare it with your answer to question Q 6.3.2.

6.4 Phase Plane for Saturating Drive

Now remember that our system is the simple one where acceleration is proportional to the input.

$$\ddot{x} = u.$$

The controlled system relies on feedback alone for damping, with the input determined by

$$u = 6(x_d - x) - 5\dot{x}$$

or

$$u = -6x - 5\dot{x}$$

if the demanded position is zero. On the lines

$$-6x - 5\dot{x} = \pm 1$$

the drive will reach its saturation value. Between these lines the phase plane plot will be exactly as we have already found it, as in Figure 6.7.

Outside the lines, the equations are completely different. The trajectories are the solutions of

$$\ddot{x} = -1$$

to the right and

$$\ddot{x} = 1$$

to the left.

To fill in the mystery region in this phase plane, we must find how the system will behave under saturated drive.

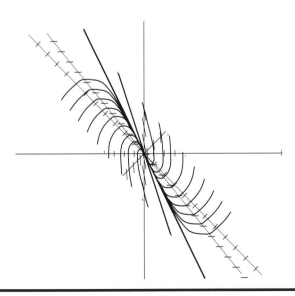

Figure 6.7   Linear region of the phase plane.

Equation 6.2 tells us that the slope of the trajectory is always given by $\ddot{x}/\dot{x}$, so in this case we are interested in the case where $\ddot{x}$ has saturated at a value of $+1$ or -1, giving the slope of the trajectories as $1/x$ or $-1/x$.

Now we see that the isoclines are no longer lines through the origin as before, but are lines of constant $\dot{x}$, parallel to the horizontal axis. If we want to find out the actual shape of the trajectories, we must solve for the relationship between x and $\dot{x}$.

On the left, where $u = 1$, we can integrate the expression for $\ddot{x}$ twice to see

$$\ddot{x} = 1$$

so

$$\dot{x} = t + a,$$

and

$$x = t^2/2 + at + b$$

from which we can deduce that

$$x = \dot{x}^2/2 + b - a^2/2,$$

i.e., the trajectories are parabolae of the form

$$x = \dot{x}^2/2 + c$$

with a horizontal axis which is the x-axis.

Similarly, the trajectories to the right, where $u = -1$, are of the form

$$x = -\dot{x}^2/2 + c.$$

The three regions of the phase plane can now be cut and pasted together to give the full picture of Figure 6.8.

We will see that the phase plane can become a powerful tool for the design of high performance position control systems. The motor drive might be proportional to the position error for small errors, but to achieve accuracy the drive must approach its limit for a small displacement. The "proportional band" is small, and for any substantial disturbance the drive will spend much of its time saturated. The ability to design feedback on a non-linear basis is then of great importance.

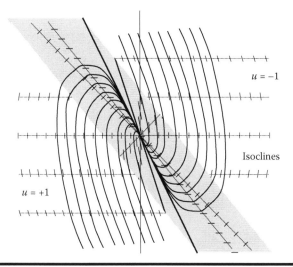

Figure 6.8 **Phase plane with linear and saturated regions.**

Q 6.4.1

The position control system is described by the same equations as before,

$$\ddot{x} + 5\dot{x} + 6x = 6w.$$

This time, however, the damping term $5\dot{x}$ is given not by feedback but by passive damping in the motor, i.e.,

$$\ddot{x} = -5\dot{x} + u$$

where

$$u = -6(x - \text{demand}).$$

Once again the drive u saturates at values $+1$ or -1. Sketch the new phase plane, noting that the saturated trajectories will no longer be parabolae. The answer can be found on the website at www.esscont.com/6/q6-4-1.htm.

Q 6.4.2

Consider the following design problem. A manufacturer of pick-and-place machines requires a robot axis. It must move a one kilogram load a distance of one meter, bringing it to rest within one second. It must hold the load at the target position with sufficient "stiffness" to resist a disturbing force, so that for a deflection of one millimeter the motor will exert its maximum restoring force.

Steps in the design are as follows:

(a) What acceleration is required to achieve the one meter movement in one second?
(b) Multiply this by an "engineering margin" of 2.5 when choosing the motor.
(c) Determine feedback values for the position and velocity signals.

You might first try pole assignment, but you would never guess the values of the poles that are necessary in practice.

Consider the "stiffness" criterion. An error of 10^{-3} meters must produce an acceleration of 10 m/s². That implies a position gain of 10,000. Use the phase plane to deduce a suitable velocity gain.

6.5 Bang–Bang Control and Sliding Mode

In the light of a saturating input, the gain can be made infinite. The proportional regional has now shrunk to a switching line and the drive always takes one limiting value or the other. This is termed *bang–bang* control. Clearly stability is now impossible, since even at the target the drive will "buzz" between the limits, but the quality of control can be excellent. Full correcting drive is applied for the slightest error.

Suppose now that we have the system

$$\ddot{x} = u,$$

where the magnitude of u is constrained to be no greater than 1.

Suppose we apply a control law in the form of logic,

```
if ((x+xdot)<0){
   u=1;
}else{
   u=-1;
}
```

then Figure 6.3 will be modified to show the same parabolic trajectories as before, with a switching line now dividing the full positive and negative drive regions. This is the line

$$x + \dot{x} = 0.$$

Now at the point (−0.5, .5), we will have $u = 1$. The slope of the trajectory will be:

$$\frac{\ddot{x}}{\dot{x}}$$

which at this point has value 2. This will take the trajectory across the switching line into the negative drive region, where the slope is now –2. And so we cross the trajectory yet again. The drive buzzes to and fro, holding the state on the switching line. This is termed *sliding*.

But the state is not glued in place. On the switching line,

$$x + \dot{x} = 0.$$

If we regard this as a differential equation, rather than the equation of a line, we see that x decays as e^{-t}.

Once sliding has started, the second order system behaves like a first order system. This principle is at the heart of *variable structure* control.

Chapter 7

Frequency Domain Methods

7.1 Introduction

The first few chapters have been concerned with the *time domain*, where the performance of the system is assessed in term of its recovery from an initial disturbance
or from a sudden change of target point. Before we move on to consider discrete
time control, we should look at some of the theory that is fundamental to stability
analysis. An assumption of linearity might lead us astray in choosing feedback
parameters for a system with drive limitation, but if the response is to settle to a
stable value the system must satisfy all the requirements for stability.

For reasons described in the Introduction, the foundations of control theory
were laid down in terms of sinusoidal signals. A sinusoidal input is applied to the
system and all the information contained in the initial transient is ignored until
the output has settled down to another sinusoid. Everything is deduced from the
"gain," the ratio of output amplitude to input, and "phase shift" of the system.

If the system has no input, testing it by frequency domain methods will be
something of a problem! But then, controlling it would be difficult, too.

There are "rules of thumb" that can be applied without questioning them, but
to get to grips with this theory, it is necessary to understand complex variables. At
the core is the way that functions can define a "mapping" from one complex plane
to another. Any course in control will always be obliged to cover these aspects,
although many engineering design tasks will benefit from a broader view.

7.2 Sine-Wave Fundamentals

Sine-waves and allied signals are the tools of the trade of classical control. They can be represented in terms of their amplitude, frequency, and phase. If you have ever struggled through a page or two of algebraic trigonometry extracting phase angles from mixtures of sines and cosines, you will realize that some computational short cuts are more than welcome. This section is concerned with the representation of sine-waves as imaginary exponentials, with a further extension to include the interpretation of complex exponentials.

DeMoivre's theorem tells us that:

$$e^{jt} = \cos(t) + j\sin(t)$$

from which we can deduce that

$$\cos(t) = (e^{jt} + e^{-jt})/2$$
$$\sin(t) = (e^{jt} - e^{-jt})/2j. \tag{7.1}$$

This can be shown by a power series expansion, but there is an alternative proof (or demonstration) much closer to the heart of a control engineer. It depends on the techniques for solving linear differential equations that were touched upon in Chapter 5. It depends also on the concept of the uniqueness of a solution that satisfies enough initial conditions.

If $y = \cos(t)$ and if we differentiate twice with respect to t, we have:

$$\dot{y} = \sin(t)$$

and

$$\ddot{y} = -\cos(t)$$

so

$$\ddot{y} = -y.$$

If we try to solve this by assuming that $y = e^{mt}$ we deduce that $m^2 = -1$, so the general solution is

$$y = Ae^{jt} + Be^{-jt}.$$

Put in the initial conditions that $\cos(0) = 1$ and $\sin(0) = 0$ and the expression follows.

7.3 Complex Amplitudes

Equation 7.1 allow us to exchange the ungainly sine and cosine functions for more manageable exponentials, but we are still faced with exponentials of both $+jt$ and $-jt$. Can this be simplified?

If we go back to

$$e^{jt} = \cos(t) + j\sin(t),$$

we can get $\cos(t)$ simply by taking the real part. If on the other hand, we take the real part of

$$(a + jb)e^{jt}$$

we get

$$a\cos(t) - b\sin(t).$$

So we can express a mixture of sines and cosines with a simple complex number. This number represents both amplitude and phase. On the strict understanding that we take the real part of any expression when describing a function of time, we can now deal in complex amplitudes of e^{jt}.

Algebra of addition and subtraction will clearly work without any complications. The real part of the sum of $(a + jb)e^{jt}$ and $(c + jd)e^{jt}$ is seen to be the sum of the individual real parts, i.e., we can add the complex numbers $(a + jb)$ and $(c + jd)$ to represent the new mixture of sines and cosines $(a + c)\cos(t) - (b + d)\sin(t)$. Beware, however, of multiplying the complex numbers to denote the product of two sine-waves. For anything of that sort you must go back to the precise representation given in Equation 7.1.

Another operation that is linear, and therefore in harmony with this representation, is differentiation. It is not hard to show that

$$\frac{d}{dt}(a + jb)e^{jt} = j(a + jb)e^{jt}.$$

Differentiation a second time or more is still a linear operation, and so each time that the mixture is differentiated we obtain an extra factor of j.

This has an enormous simplifying effect on the task of solving differential equations for steady solutions with sinusoidal forcing functions. In the "knife and fork" approach we would have to assume a result of the form $A\cos(t) + B\sin(t)$, substitute this into the equations and unscramble the resulting mess of sines and cosines. Let us use an example to see the improvement.

Q 7.3.1

The system described by the second order differential equation

$$\ddot{x} + 4\dot{x} + 5x = u$$

is forced by the function

$$u = 2\cos(3t) + \sin(3t).$$

What is the steady state solution (after the effects of any initial transients have died away)?

The solution will be a mixture of sines and cosines of $3t$, which we can represent as the real part of Xe^{3jt}.

The derivative of x will be the real part of $3jXe^{3jt}$.

When we differentiate a second time, we multiply this by another $3j$, so the second derivative will be the real part of $-9Xe^{3jt}$.

At the same time, u can be represented as the real part of $(2 - j)e^{3jt}$.

Substituting all of these into the differential equation produces a bundle of multiples of e^{3jt}, and so we can take the exponential term out of the equation and just equate the coefficients. We get:

$$(-9)X + 4(3j)X + 5X = (2 - j)$$

i.e.,

$$(-4 + 12j)X = (2 - j)$$

and so

$$X = \frac{2 - j}{-4 + 12j} = \frac{(2 - j)(-4 - 12j)}{4^2 + 12^2} = \frac{4 - 20j}{160}$$

$$X = 1/40 - j/8.$$

The final solution is

$$x = -\frac{1}{40}\cos(3t) + \frac{1}{8}\sin(3t).$$

As a masochistic exercise, solve the equation again the hard way, without using complex notation.

7.4 More Complex Still-Complex Frequencies

We have tried out the use of complex numbers to represent the amplitudes and phases of sinusoids. Could we usefully consider complex frequencies too? Yes, anything goes.

How can we interpret $e^{(\lambda + j\omega)t}$? Well it immediately expands to give $e^{\lambda t}e^{j\omega t}$, in other words the imaginary exponential that we now know as a sinusoid is multiplied by a real exponential of time. If the value of λ is negative, then the envelope of the sinusoid will decay toward zero, rather like the clang of a bell. If λ is positive, however, the amplitude will swell indefinitely.

Now we can represent the value of $\lambda + j\omega$ by a point in a plane where λ is plotted horizontally and ω is vertical. We can illustrate the functions of time represented by points in this "frequency plane" by plotting them in an array of small panels, as in Figure 7.1.

Code at www.esscont.com/7/responses.htm and www.esscont.com/7/responses2.htm has been used to produce this figure.

The signal shown is plotted over three seconds. The center column represents "pure" frequencies where λ is zero. The middle row is for zero frequency, where the function is just an exponential function of time. You will see that the scale has been compressed so that λ has a range of just ± 1, while the frequencies run from -4 to 4. But these are angular frequencies of radians per second, where 2π radians per second correspond to one cycle per second.

The process of differentiation is still linear, and we see that:

$$\frac{d}{dt}(a + jb)e^{(\lambda + j\omega)t} = (a + jb)(\lambda + j\omega)e^{(\lambda + j\omega)t}.$$

In other words, we can consider forcing functions that are products of sinusoids and exponentials, and can take exactly the same algebraic short cuts as before.

Q 7.4.1

Consider the example Q 7.3.1 above, when the forcing function is instead $e^{-t}\cos(2t)$.

Q 7.4.2

Consider Q 7.3.1 again with an input function $e^{-2t}\sin(t)$. What went wrong? Read on to find out.

7.5 Eigenfunctions and Gain

Classical control theory is concerned with linear systems. That is to say, the differential equations contain constants, variables, and their derivatives of various orders, but never the products of variables, whether states or inputs.

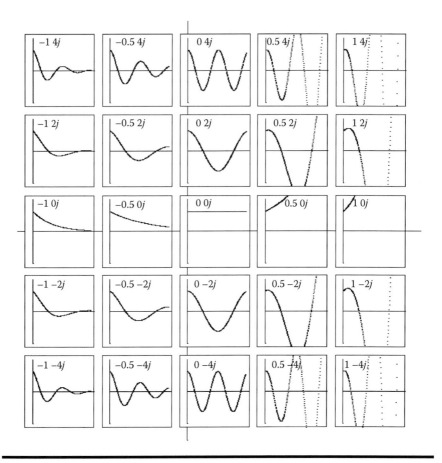

Figure 7.1   Complex frequencies.

Now the derivative of a sine-wave is another sine-wave (or cosine-wave) of the same frequency, shifted in phase and probably changed in amplitude. The sum of two sine-waves of the same frequency and assorted phases will be yet another sine-wave of the same frequency, with phase and amplitude which can be found by a little algebra. No matter how many derivatives we add, if the basic signal is sinusoidal then the mixture will also be sinusoidal.

If we apply a sinusoidal input to a linear system, allowing time for transients to die away, the output will settle down to a similar sinusoid. If we double the size of the input, the output will also settle to double its amplitude, while the phase relationship between input and output will remain the same. The passage of the sine-wave through the system will be characterized by a "gain," the ratio between output and input magnitude, and a phase shift.

In the previous sections of this chapter, we saw that a phase shift can be represented by means of a complex value of gain, when the sine-wave is expressed in its complex exponential form $e^{j\omega t}$. Now we see that applying such a signal to the input of the linear system will produce exactly the same sort of output, multiplied only by a constant gain (probably complex). The signal $e^{j\omega t}$ is an "eigenfunction" of any linear system.

Altering the frequency of the input signal will of course change the gain of the response; the gain is a function of the test frequency. If tests are made at a variety of frequencies, then a curve can be plotted of gain against frequency, the "frequency response." As we will see, a simple series of such tests can in most cases give all the information needed to design a stable controller.

The sine-wave is not alone in the role of eigenfunction. Any exponential e^{st} will have similar properties. For an input of this form and for the correct initial conditions, the system will give an output $G(s)e^{st}$, where $G(s)$ is the gain for that particular value of the constant s. Clearly if s is real, e^{st} is a less convenient test signal to use than a sine-wave. If s is positive, the signal will grow to enormous proportions. If s is negative, the experimenter will have to be swift to catch the response before it dies away to nothing. Although of little use in experimental terms, the mathematical significance of these more general signals is very important, especially when s is allowed to be complex.

7.6 A Surfeit of Feedback

In the case of a position servo, just as for an amplifier, it is natural to seek as high a "loop gain" as possible. We want the correction motor to exert a large torque for as small an error as possible. When the dynamics are accurately known in state-space form, feedback can be determined by analytical considerations. In the absence of such insight, early engineers had to devise practical methods of finding the limit to which simple feedback gain could be increased. These are, of course, still of great practical use today.

As was introduced in Chapter 1, feedback has many more roles in electronics, particularly for reducing the non-linearity of amplifiers, and for reducing uncertainty in their gains. An amplifier stage might have a gain, say, with value between 50 and 200. Two such stages could give a range of combined gains between 2500 and 40,000—a factor of 16. Suppose we require an accurate target gain of 100, how closely can we achieve it?

We apply feedback by mixing a proportion k of the output signal with the input. To work out the resulting gain, it is easiest to work backward. To obtain an output of one volt, the input to the amplifier must be $1/G$ volts, where G is the gain of the open loop amplifier. We now feed back a further proportion k of the output, in such a sense as to make the necessary input greater. (Negative feedback.)

Now we have:

$$v_{in} = \left(\frac{1}{G} + k\right) v_{out}$$

which can be rearranged to give the gain, the ratio of output to input, as:

$$\frac{G}{1 + kG}.$$

To check up on the accuracy of the resulting gain, this can be rearranged as:

$$\frac{1/k}{1 + 1/(kG)}.$$

Now if we are looking for a gain of 100, then $k = 0.01$. If G lies between 2500 and 40,000, then kG is somewhere between 25 and 400. We see that the closed loop gain may be just 4% below target if G takes the lowest value. The uncertainty in gain has been reduced dramatically. The larger the minimum value of kG, the smaller will be the uncertainty in gain, and so a very large loop gain would seem to be desirable.

Of course, positive feedback will have a very different effect. The feedback now assists the input, increasing the closed loop gain to a value:

$$\frac{G}{1 - kG}.$$

If k starts at a very small value and is progressively increased, something dramatic happens when $kG = 1$. Here the closed loop gain becomes infinite; and the output can flip from one extreme to the other at the slightest provocation. If k is increased further, the system becomes "bistable," giving an output at one extreme limit until an input opposes the feedback sufficiently to flip it to the other extreme.

All would be well with using huge negative-feedback loop gains if the amplifier responded infinitely rapidly, but unfortunately it will contain some dynamics. The open loop gain is not a constant G, but is seen to be a function of the applied test frequency $G(j\omega)$, complete with phase shift. In any but the simplest model of the amplifier, this phase shift can approach or reach 180°, and that is where trouble can break out. A phase shift of 180° is equivalent to a reversal in sign of the original sine-wave. Negative feedback becomes positive, and if the value of kG still has magnitude greater than unity, then closing the loop will certainly result in oscillation.

The determination of a permissible level of k will depend on the race between increasing phase shift and diminishing gain as the test frequency is increased. We could measure the phase shift at the frequency where kG just falls below unity. As long as this is below 180°, we have some margin of safety—the actual shortfall is called the "phase margin." Alternatively, we could measure the gain at the frequency that gives just 180° phase shift. We call the amount by which kG falls below unity the "gain margin." In the early days, these led to rules of thumb, then they became an art, and now a science. We can put the methods onto a firm foundation of mathematics.

7.7 Poles and Polynomials

Analysis of the servomotor problem from the state-space point of view gave us a list of first order differential equations. A "lumped linear system" of this type will have a set of state equations where each has a simple d/dt on the left and a linear combination of state variables and inputs on the right. There are of course many other systems that fall outside such a description, but why look for trouble.

For a start, let us suppose that the system has a single input and a single output:

$$\dot{\mathbf{x}} = \mathbf{A}\mathbf{x} + \mathbf{B}u$$

and

$$y = \mathbf{C}\mathbf{x}.$$

With a certain amount of algebraic juggling, we can eliminate all the x's from these equations, and get back again to the "traditional" form of a single higher order equation linking input and output. This will be of the form:

$$\frac{d^n y}{dt^n} + a_1 \frac{d^{n-1} y}{dt^{n-1}} + a_2 \frac{d^{n-2} y}{dt^{n-2}} + \cdots + a_n y = b_0 \frac{d^m u}{dt^m} + b_1 \frac{d^{m-1} u}{dt^{m-1}} + \cdots + b_m u. \quad (7.2)$$

Now let us try the system out with an input that is an exponential function of time e^{st}—without committing ourselves to stating whether s is real or complex. If we assume that the initial transients have all died away, then y will also be proportional to the same function of time. Since the derivative of e^{st} is the same function, but multiplied by s, all the time derivatives simply turn into powers of s. We end up with:

$$(s^n + a_1 s^{n-1} + a_2 s^{n-2} + \cdots + a_n) y = (b_0 s^m + b_1 s^{m-1} + \cdots + b_m) u. \quad (7.3)$$

The gain, the ratio between output and input, is now the ratio of these two polynomials in s.

$$G(s) = \frac{b_0 s^m + b_1 s^{m-1} + \cdots + b_m}{s^n + a_1 s^{n-1} + a_2 s^{n-2} + \cdots + a_n}.$$

If we commit ourselves to making s be pure imaginary, with value $j\omega$, we obtain an expression for the gain (and phase shift) at any frequency.

Now any polynomial can be factorized into a product of linear terms of the form:

$$(s - p_1)(s - p_2)...(s - p_n),$$

where the coefficients are allowed to be complex. Clearly, if s takes the value p_1, then the value of the polynomial will become zero. But what if the polynomial in question is the denominator of the expression for the gain? Does it not mean that the gain at complex frequency p_1 is infinite? Yes, it does.

The gain function, in the form of the ratio of two polynomials in s, is more commonly referred to as the "transfer function" of the system, and the values of s that make the denominator zero are termed its "poles."

It is true that the transfer function becomes infinite when s takes the values of one of the poles, but this can be interpreted in a less dramatic way. The ratio of output to input can just as easily be infinite when the input is zero for a non-zero output. In other words, we can get an output of the form $e^{p_i t}$ for no input at all, where p_i is any pole of the transfer function.

Now if the pole has a real, negative value, say -5, it means that there can be an output e^{-5t}. This is a rapidly decaying transient, which might have been provoked by some input before we set the input to zero. This sort of transient is unlikely to cause any problem.

Suppose instead that the pole has value $-1 + j$. The function $e^{(-1 + j)t}$ can be factorized into $e^{-t} e^{jt}$. Clearly it represents the product of a cosine-wave of angular frequency unity with a decaying exponential. After an initial "ping" the response will soon cease to have any appreciable value, all is still well.

Now let us consider a pole that is purely imaginary, $-2j$, say. The response e^{-2jt} never dies away. We are in trouble.

Even worse, consider a pole at $+1 + j$. Now we have a sine-wave multiplied by an exponential which more than doubles each second. The system is hopelessly unstable.

We conclude that poles that have negative real parts are relatively benign, causing no trouble, but poles which have a real part that is positive, or even zero, are a sign of instability. What is more, even one such pole among a host of stable ones is enough to make a system unstable.

For now, let us see how this new insight helped the traditional methods of examining a system.

7.8 Complex Manipulations

The logarithm of a product is the sum of the individual logarithms. If we take the logarithm of the gain of a system described by a ratio of polynomials, we are left adding and subtracting logarithms of expressions no more complicated than $(s - p_i)$, the factors of the numerator or denominator. To be more precise, if

$$G(s) = \frac{(s - z_1)(s - z_2)...(s - z_m)}{(s - p_1)(s - p_2)...(s - p_n)}$$

then,

$$\ln(G(s)) = \ln(s - z_1) + \ln(s - z_2) + \cdots + \ln(s - z_m)$$
$$- \ln(s - p_1) - \ln(s - p_2) - \cdots - \ln(s - p_n),$$

where $\ln(G)$ is the "natural" logarithm to base e.

First of all, we are likely to want to work out a frequency response, by substituting the value $j\omega$ for s, and we are faced with a set of logarithms of complex expressions.

Now a complex number can be expressed in polar form as $re^{j\theta}$ as in Figure 7.2. (Remember that $e^{j\theta} = \cos\theta + j\sin\theta$.) Here r is the modulus of the number, the square root of the sum of the squares of real and imaginary parts, while θ is the "argument," an angle in radians whose tangent gives the ratio of imaginary to real parts. When we take the logarithm of this product, we see that it splits neatly into a real part, $\ln(r)$, and an imaginary part, $j\theta$.

Let us consider a system with just one pole, with gain

$$G(s) = \frac{1}{s - p}.$$

(Note that for stability, p will have to be negative.) Substitute $j\omega$ for s and we find:

$$\ln(G(j\omega)) = -\ln(j\omega - p)$$
$$= -\ln(\sqrt{\omega^2 + p^2}) + j\tan^{-1}\frac{\omega}{p}.$$

The real part is concerned with the magnitude of the output, while the imaginary part determines the phase. In the early days of electronic amplifiers, phase was hard to measure. The properties of the system had to be deduced from the amplitude alone.

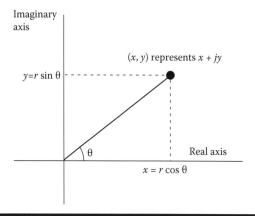

Figure 7.2   Illustration of polar coordinates.

Clearly, for very low frequencies the gain will approximate to $1/-p$, i.e., the log gain will be roughly constant. For very high frequencies, the term ω^2 will dominate the expression in the square root, and so the real part of log gain will be approximately

$$-\ln(\omega).$$

If we plot $\ln(\text{gain})$ against $\ln(\omega)$, we will obtain a line of the form $y = -x$, i.e., a slope of -1 passing through the origin. A closer look will show that the horizontal line that roughly represents the low frequency gain meets this new line at the value $\omega = -p$. Now these lines, although a fair approximation, do not accurately represent the gain. How far are they in error? Well if we substitute the value p^2 for ω^2 the square root will become $\sqrt{2}|p|$, i.e., the logarithm giving the real part of the gain will be:

$$-\ln(|p|) - \ln(\sqrt{2}).$$

The gain at a frequency equal to the magnitude of the pole is thus a factor $\sqrt{2}$ less than the low frequency gain. Plot a frequency response, and this "breakpoint" will give away the value of the pole. In Figure 7.3 is a sketch showing the frequency response when p has the value -2.

7.9 Decibels and Octaves

Let us briefly turn aside to us settle some of the traditional terminology you might come across, which could prove confusing.

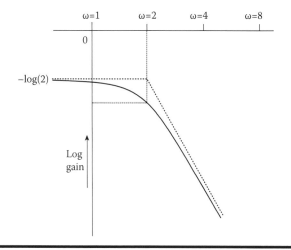

Figure 7.3   Sketch of Bode plot for 1/(s + 2).

Remember that in taking frequency responses, the early engineers were concerned with telephones. They measured the output of the system not by its amplitude but by the power of its sound. This was measured on a logarithmic scale, but the logarithm base was 10. Ten times the output power was one "bel." Now a factor of $\sqrt{2}$ in amplitude gives a factor of two in power, and is thus $\log_{10}(2)$ "bels," or around 0.3 bel. The bel is clearly rather a coarse unit, so we might redefine this as 3 "decibels." The "breakpoint" is found when the gain is "three decibels down."

We have the gain measured on a logarithmic scale, even if the units are a little strange. Now, for the frequency. Musicians already measure frequency on a logarithmic scale, but the semitone does not really appeal to an engineer as a unit of measurement. Between one "C" on the piano and the next one higher, the frequency is doubled. The unit of log frequency used by the old engineers was therefore the "octave," a factor of two.

Now we can plot log power in decibels against log frequency in octaves. What has become of the neat slope of −1 we found above? At high frequencies, the amplitude halves if the frequency is doubled. The power therefore drops by four, giving a fall of 6 decibels for each octave. Keep the slogans "three decibels down" and "six decibels per octave" safe in your memory!

7.10 Frequency Plots and Compensators

Let us return to simpler units, and look again at the example of

$$G(s) = \frac{1}{s+2}.$$

We have noted that the low frequency gain is close to 1/2, while the high frequency gain is close to 1/ω. We have also seen that at ω = 2 the gain has fallen by a factor of √2. Note that at this frequency, the real and imaginary parts of the denominator have the same magnitude, and so the phase shift is a lag of 45°—or in radian terms π/4. As the frequency is increased, the phase lag increases toward 90°.

Now we can justify our obsession with logarithms by throwing in a second pole, let us say:

$$G(s) = \frac{10}{(s+2)(s+5)}.$$

The numerator of 10 will keep up our low frequency gain to unity. Now we can consider the logarithm of this gain as the sum of the two logarithms

$$\ln(G(s)) = \ln\left(\frac{2}{s+2}\right) + \ln\left(\frac{5}{s+5}\right).$$

The first logarithm is roughly a horizontal line at value zero, diving down at a slope of –1 from a breakpoint at ω = 2. The second is similar, but with a breakpoint at ω = 5. Put them together, and we have a horizontal line, diving down at ω = 2 with a slope of –1, taking a further nosedive at ω = 5 with a slope of –2. If we add the phase shifts together, the imaginary parts of the logarithmic expressions, we get the following result, illustrated in Figure 7.4.

At low frequency, the phase shift is nearly zero. As the frequency reaches 2 radians per second, the phase shift has increased to 45°. As we increase frequency beyond the first pole, its contribution approaches 90° while the second pole starts to take effect. At ω = 5, the phase shift is around 135°. As the frequency increases further, the phase shift approaches 180°. It never "quite" reaches it, so in theory we could never make this system unstable, however, much feedback we applied. We could have a nasty case of resonance, however, but only at a frequency well above 5 radians per second.

In this simple case, we can see a relationship between the slope of the gain curve and the phase shift. If the slope of the "tail" of the gain is –1, the ultimate phase shift is 90°—no problem. If the slope is –2, we might be troubled by a resonance. If it is –3, the phase shift is heading well beyond 180° and we must be wary. Watch out for the watershed at "12 decibels per octave."

In this system, there are no phase shift effects that cannot be predicted from the gain. That is not always the case. Veteran engineers lay in dread of "non-minimum-phase" systems.

Consider the system defined by

$$G(j\omega) = \frac{jw - 2}{jw + 2}.$$

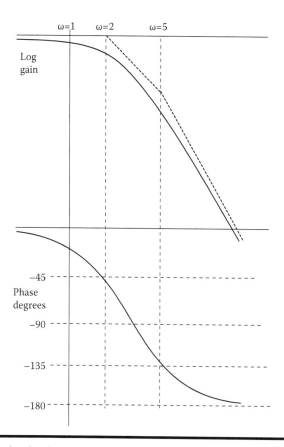

Figure 7.4   Bode plot for 10/(s + 2)(s + 5).

Look at the gain, and you will see that the magnitude of the numerator is equal to that of the denominator. The gain is unity at all frequencies. The phase shift is a different matter entirely. At low frequencies there is an inversion—a lead of 180°—with phase lead reducing via 90° at 2 radians per second to no shift at high frequencies. A treacherous system to consider for the application of feedback! This is a classic example of a non-minimum-phase system. (In general, a system is non-minimum phase if it has zeros in the right half of the complex frequency plane.)

The plot of log amplitude against log frequency is referred to as a Bode diagram. It appears that we can add together a kit of breakpoints to predict a response, or alternatively inspect the frequency response to get an insight into the transfer function. On the whole this is true. However, any but the simplest systems will require considerable skill to interpret.

The Bode plot which includes a second curve showing phase shift is particularly useful for suggesting the means of squeezing a little more loop gain, or for stabilizing an awkward system. If the feedback is not a simple proportion of the output, but instead contains some "compensator" with gain function $F(s)$, then the stability will be dictated by the product of the two gains, $F(s)G(s)$.

The closed loop gain will be

$$\frac{G(s)}{1 + F(s)G(s)}$$

and so the poles of the closed loop system are the roots of

$$1 + F(s)G(s) = 0.$$

A "phase advance" circuit can be added into the loop of the form:

$$F(j\omega) = \frac{3j\omega + a}{j\omega + 3a}.$$

This particular compensator will have a low frequency gain of one-third and a high frequency gain of three. It might therefore appear to whittle down the gain margin. At a second glance, however, it is seen to give a positive phase shift. In this case the phase shift reaches $\tan^{-1}(4/3)$ at frequency a, enabling an awkward second order system to be tamed without a resonance.

Q 7.10.1

Suppose that the system to be controlled is an undamped motor, appearing as two integrators in cascade. Now $G(s) = 1/s^2$. The phase shift starts off at $180°$, so any proportional position feedback will result in oscillation. By adding the above phase-advance circuit into the loop, a response can be obtained with three equal real roots. As an exercise, solve the algebra to derive their value. Bode plots of the phase-advance compensator and of the combined system are shown in Figures 7.5 and 7.6.

7.11 Second Order Responses

So far we have considered only real poles and zeros. They can also come in complex conjugate pairs, and we should have a look at the result before moving on. Consider

$$G(s) = \frac{a^2}{s^2 + kas + a^2}.$$

Now, the denominator will factorize into the product of two terms,

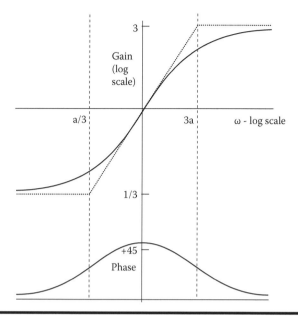

Figure 7.5   Bode plot of phase-advance $(a + 3j\omega)/(3a + j\omega)$.

$$(s - p_1)(s - p_2).$$

If k has value of 2 or greater, p_1 and p_2 are real. If k is less than 2, then the roots will form a conjugate complex pair.

If we work out the frequency response by substituting $j\omega$ for s, then we see that when $\omega = a$ the real parts of the denominator cancel out. At that frequency the phase shift is 90°. We have

$$G(ja) = \frac{1}{kj} = \frac{-j}{k}.$$

If k is steadily reduced, an increasing peak in the response is seen, tending toward infinite amplitude as k tends to zero. The responses are shown on a logarithmic scale in Figure 7.7.

7.12 Excited Poles

Exercise Q 7.4.2 earlier in this chapter seemed to require an answer that resulted from an infinite gain. A system was excited at a complex frequency corresponding to a pole of the gain function. Should its output really be infinite?

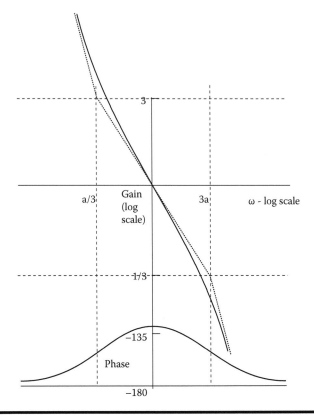

Figure 7.6 Bode plot showing stabilization of a^2/s^2 using phase-advance $(a + 3j\omega)/(3a + j\omega)$.

Consider a simpler example, the system described by gain

$$G(s) = \frac{1}{s+a},$$

when the input is e^{-at}, i.e., $s = -a$.

If we turn back to the differential equation that the gain function represents, we see

$$\dot{x} + ax = e^{-at}.$$

Since the "complementary function" is now the same as the input function, we must look for a "particular integral" which has an extra factor of t, i.e., the general solution is:

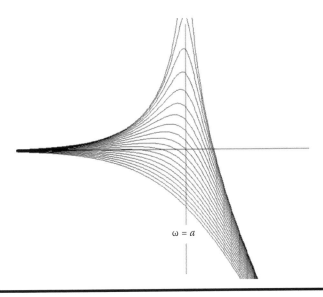

$\omega = a$

Figure 7.7 Family of response curves for various damping factors. (Screen grab from www.esscont.com/7/7-7-damping.htm)

$$x = te^{-at} + Ae^{-at}.$$

As t becomes large, we see that the ratio of output to input also becomes large—but the output still tends rapidly to zero.

Even if the system had a pair of poles representing an undamped oscillation, applying the same frequency at the input would only cause the amplitude to ramp upwards at a steady rate; and there would be no sudden infinite output. Let one of the poles stray so that its real part becomes positive, however, and there will be an exponential runaway in the amplitude of the output.

Chapter 8

Discrete Time Systems and Computer Control

8.1 Introduction

There is a belief that discrete time control is somehow more difficult to understand than continuous control. It is true that a complicated approach can be made to the subject, but the very fact that we have already considered digital simulation shows that there need be no hidden mysteries.

The concept of differentiation, with its need to take limits of ratios of small increments that tend to zero, is surely more challenging than considering a sequence of discrete values. But first let us look at discrete time control in general.

When dedicated to a real-time control task, a computer measures a number of system variables, computes a control action, and applies a corrective input to the system. It does not do this continuously, but at discrete instants of time. Some processes need frequent correction, such as the attitude of an aircraft, while on the other hand the pumps and levels of a sewage process might only need attention every five minutes.

Provided the corrective action is sufficiently frequent, there seems on the surface to be no reason for insisting that the intervals should be regular. When we look deeper into the analysis, however, we will find that we can take shortcuts in the mathematics if the system is updated at regular intervals. We find ourselves dealing with difference equations that have much in common with the methods we can use for differential equations.

Since we started with continuous state equations, we should start by relating these to the discrete time behavior.

8.2 State Transition

We have become used to representing a linear continuous system by the state equations:

$$\dot{x} = Ax + Bu. \tag{8.1}$$

Now, let us say that the input of our system is driven by a digital-to-analog converter, controlled by the output of a computer. A value of input **u** is set up at time t and remains constant until the next output cycle at $t + T$.

If we sample everything at constant intervals of length T and write $t = nT$, we find that the equivalent discrete time equations are of the form:

$$x_{n+1} = Mx_n + Nu_n,$$

where x_n denotes the state measured at the nth sampling interval, at $t = nT$.

By way of a proof (which you can skip if you like, going on to Section 8.3) we consider the following question. If x_n is the state at time t, what value will it reach at time $t + T$?

In the working that follows, we will at first simplify matters by taking the initial time, t, to be zero.

In Section 2.2, we considered the solution of a first order equation by the "integrating factor" method. We can use a similar technique to solve the multivariable matrix equation, provided we can find a matrix e^{-At} whose derivative is $e^{-At}(-A)$.

An exponential function of a matrix might seem rather strange, but it becomes simple when we consider the power series expansion. For the scalar case we had:

$$e^{at} = 1 + at + a^2 \frac{t^2}{2!} + a^3 \frac{t^3}{3!} + \cdots$$

When we differentiate term by term we see that

$$\frac{d}{dt} e^{at} = 0 + a + a^2 t + a^3 \frac{t^2}{2!} + \cdots$$

and when we compare the series we see that each power of t in the second series has an extra factor of a. From the series definition it is clear that

$$\frac{d}{dt} e^{at} = e^{at} a.$$

The product $\mathbf{A}t$ is simply obtained by multiplying each of the coefficients of $\mathbf{A}$ by t.

For the exponential, we can define

$$e^{\mathbf{A}t} = \mathbf{I} + \mathbf{A}t + \mathbf{A}^2 \frac{t^2}{2!} + \mathbf{A}^3 \frac{t^3}{3!} + \cdots$$

where $\mathbf{I}$ is the unit matrix. Now,

$$\frac{d}{dt} e^{\mathbf{A}t} = 0 + \mathbf{A} + \mathbf{A}^2 t + \mathbf{A}^3 \frac{t^2}{2!} + \cdots$$

so just as in the scalar case we have

$$\frac{d}{dt} e^{\mathbf{A}t} = e^{\mathbf{A}t} \mathbf{A}$$

By the same token,

$$\frac{d}{dt} e^{-\mathbf{A}t} = -e^{-\mathbf{A}t} \mathbf{A}.$$

There is a good reason to write the $\mathbf{A}$ matrix after the exponential.

The state equations, 8.1, tell us that:

$$\dot{\mathbf{x}} = \mathbf{A}\mathbf{x} + \mathbf{B}\mathbf{u}$$

so

$$\dot{\mathbf{x}} - \mathbf{A}\mathbf{x} = \mathbf{B}\mathbf{u}$$

and multiplying through by $e^{-\mathbf{A}t}$ we have

$$e^{-\mathbf{A}t} \dot{\mathbf{x}} - e^{-\mathbf{A}t} \mathbf{A}\mathbf{x} = e^{-\mathbf{A}t} \mathbf{B}\mathbf{u}$$

and just as in the scalar case, the left-hand side can be expressed as the derivative of a product

$$\frac{d}{dt} (e^{-\mathbf{A}t} \mathbf{x}) = e^{-\mathbf{A}t} \mathbf{B}\mathbf{u}.$$

Integrating, we see that

$$\left[e^{-\mathbf{A}t}\mathbf{x}\right]_0^T = \int_0^T e^{-\mathbf{A}t}\mathbf{B}u\,dt.$$

When $t = 0$, the matrix exponential is simply the unit matrix, so

$$\left[e^{-\mathbf{A}T}\mathbf{x}(T) - \mathbf{x}(0)\right] = \int_0^T e^{-\mathbf{A}t}\mathbf{B}u\,dt$$

and we can multiply through by $e^{\mathbf{A}t}$ to get

$$\mathbf{x}(T) - e^{\mathbf{A}T}\mathbf{x}(0) = e^{\mathbf{A}T}\int_0^T e^{-\mathbf{A}t}\mathbf{B}u\,dt$$

which can be rearranged as

$$\mathbf{x}(T) = e^{\mathbf{A}T}\mathbf{x}(0) + e^{\mathbf{A}T}\int_0^T e^{-\mathbf{A}t}\mathbf{B}u\,dt.$$

What does this mean?

Since $\mathbf{u}$ will be constant throughout the integral, the right-hand side can be rearranged to give

$$\mathbf{x}(T) = e^{\mathbf{A}T}\mathbf{x}(0) + \left\{ e^{\mathbf{A}T}\int_0^T e^{-\mathbf{A}t}\mathbf{B}\,dt \right\}\mathbf{u}(0)$$

which is of the form:

$$\mathbf{x}(T) = \mathbf{M}\mathbf{x}(0) + \mathbf{N}\mathbf{u}(0)$$

where **M** and **N** are constant matrices once we have given a value to T. From this it follows that

$$\mathbf{x}(t+T) = \mathbf{Mx}(t) + \mathbf{Nu}(t)$$

and when we write nT for t, we arrive at the form:

$$\mathbf{x}_{n+1} = \mathbf{Mx}_n + \mathbf{Nu}_n$$

where the matrices **M** and **N** are calculated from

$$\mathbf{M} = e^{\mathbf{A}T}$$

and

$$\mathbf{N} = e^{\mathbf{A}T} \int_0^T e^{-\mathbf{A}t} \mathbf{B} dt$$

while the system output is still given by

$$\mathbf{y}_n = \mathbf{Cx}_n.$$

The matrix $\mathbf{M} = e^{\mathbf{A}T}$ is termed the "state transition matrix."

8.3 Discrete Time State Equations and Feedback

As long as there is a risk of confusion between the matrices of the discrete state equations and those of the continuous ones, we will use the notation **M** and **N**. (Some authors use **A** and **B** in both cases, although the matrices have different values.)

Now if our computer is to provide some feedback control action, this must be based on measuring the system output, $\mathbf{y}_n$, taking into account a command input, $\mathbf{v}_n$, and computing an input value $\mathbf{u}_n$ with which to drive the digital-to-analog converters. For now we will assume that the computation is performed instantaneously as far as the system is concerned, i.e., the intervals are much longer than the computing time. We see that if the action is linear,

$$\mathbf{u}_n = \mathbf{Fy}_n + \mathbf{Gv}_n \tag{8.2}$$

where $\mathbf{v}_n$ is a command input.

As in the continuous case, we can substitute the expression for $\mathbf{u}$ back into the system equations to get

$$\mathbf{x}_{n+1} = \mathbf{M}\mathbf{x}_n + \mathbf{N}(\mathbf{F}\mathbf{y}_n + \mathbf{G}\mathbf{v}_n)$$

and since $\mathbf{y}_n = \mathbf{C}\mathbf{x}_n$,

$$\mathbf{x}_{n+1} = (\mathbf{M} + \mathbf{N}\mathbf{F}\mathbf{C})\mathbf{x}_n + \mathbf{N}\mathbf{G}\mathbf{v}_n. \tag{8.3}$$

Exactly as in the continuous case, we see that the system matrix has been modified by feedback to describe a different performance. Just as before, we wish to know how to ensure that the feedback changes the performance to represent a "better" system. But to do this, we need to know how to assess the new state transition matrix $\mathbf{M} + \mathbf{N}\mathbf{F}\mathbf{C}$.

8.4 Solving Discrete Time Equations

When we had a differential equation like

$$\ddot{x} + 5\dot{x} + 6x = 0,$$

we looked for a solution in the form of e^{mt}.

Suppose that we have the difference equation

$$x_{n+2} + 5x_{n+1} + 6x_n = 0$$

what "eigenfunction" can we look for?

We simply try $x_n = k^n$ and the equation becomes

$$k^{n+2} + 5k^{n+1} + k^n = 0,$$

so we have

$$(k^2 + 5k + 6)k^n = 0$$

and once more we find ourselves solving a quadratic to find $k = -2$ or $k = -3$.

The roots are in the left-hand half plane, so will this represent a stable system? Not a bit!

From an initial value of one, the sequence of values for x can be

1, –3, 9, –27, 81…

So what is the criterion for stability? x must die away from any initial value, so $|k| < 1$ for all the roots of any such equation. In the cases where the roots are complex it is the size, the modulus of k that has to be less than one. If we plot the roots in the complex plane, as we did for the frequency domain, we will see that the roots must lie within the "unit circle."

8.5 Matrices and Eigenvectors

When we multiply a vector by a matrix, we get another vector, probably of a different magnitude and in a different direction. Just suppose, though, that for a special vector the direction was unchanged. Suppose that the new vector was just the old vector, multiplied by a scalar constant k. Such a vector would be called an "eigenvector" of the matrix and the constant k would be the corresponding "eigenvalue."

If we repeatedly multiply that vector by the matrix, doing it n times, say, then the resulting vector will be k^n times the original vector. But that is just what happens with our state transition matrix. If we keep the command input at zero, the state will be repeatedly multiplied by $\mathbf{M}$ as each interval advances. If our initial state coincides with an eigenvector, we have a formula for every future value just by multiplying every component by k^n.

So how can we find the eigenvectors of $\mathbf{M}$?

Suppose that ξ is an eigenvector. Then

$$\mathbf{M}\xi = k\xi = k\mathbf{I}\xi.$$

We can move everything to the left to get

$$\mathbf{M}\xi - k\mathbf{I}\xi = 0,$$

$$(\mathbf{M} - k\mathbf{I})\xi = 0.$$

Now, the zero on the right is a vector zero, with all its components zero. Here we have a healthy vector, ξ, which when multiplied by $(\mathbf{M} - k\mathbf{I})$ is reduced to a vector of zeros.

One way of viewing a product such as $\mathbf{Ax}$ is that each column of $\mathbf{A}$ is multiplied by the corresponding component of $\mathbf{x}$ and the resulting vectors added together. We know that in evaluating the determinant of **A**, we can add combinations of columns together without changing the determinant's value. If we can get a resulting column that is all zeros, then the determinant must clearly be zero too.

So in this case, we see that

$$\det(\mathbf{M} - k\mathbf{I}) = 0.$$

This will give us a polynomial in k of the same order as the dimension of **M**, with roots k_1, k_2, k_3...

When we substitute each of the roots, or "eigenvalues" back into the equation, we can solve to find the corresponding eigenvector. With a set of eigenvectors, we can even devise a transformation of axes so that our state is represented as the sum of multiples of them.

If we have initial

$$\mathbf{X}_0 = a_1\xi_1 + a_2\xi_2 + a_3\xi_3 \ldots$$

then at the nth interval we will have

$$\mathbf{X}_n = a_1 k_1^n \xi_1 + a_2 k_2^n \xi_2 + a_3 k_3^n \xi_3 \ldots$$

8.6 Eigenvalues and Continuous Time Equations

Eigenvalues can help us find a solution to the discrete time equations. Can they help out too in the continuous case?

To find the response to a transient, we must solve

$$\dot{\mathbf{x}} = \mathbf{A}\mathbf{x},$$

since we hold the input at zero.

Now if **x** is proportional to one of the eigenvectors of **A**, we will have

$$\dot{\mathbf{x}} = \lambda\mathbf{x}$$

where λ is the corresponding eigenvalue. And by now we have no difficulty in recognizing the solution as

$$\mathbf{x}(t) = e^{\lambda t}\mathbf{x}(0).$$

So if a general state vector **x** has been broken down into the sum of multiples of eigenvectors

$$\mathbf{x}(0) = a_1 \xi_1 + a_2 \xi_2 + a_3 \xi_3 \ldots$$

then

$$\mathbf{x}(t) = a_1 e^{\lambda_1 t} \xi_1 + a_2 e^{\lambda_2 t} \xi_2 + a_3 e^{\lambda_3 t} \xi_3 \ldots$$

Cast your mind back to Section 5.5, where we were able to find variables w_1 and w_2 that resulted in a diagonal state equation. Those were the eigenvectors of that particular system.

8.7 Simulation of a Discrete Time System

We were able to simulate the continuous state equations on a digital computer, and by taking a small time-step we could achieve a close approximation to the system behavior. But now that the state equations are discrete time, we can simulate them exactly. There is no approximation and part of the system may itself represent a computer program. In the code which follows, remember that the index in brackets is not the sample number, but the index of the particular component of the state or input vector. The sample time is "now."

Assume that all variables have been declared and initialized, and that we are concerned with computing the next value of the state knowing the input **u**. The state has *n* components and there are *m* components of input. For the coefficients of the discrete state matrix, we will use A[i][j], since there is no risk of confusion, and we will use B[i][j] for the input matrix:

```
for(i=1;i<=n;i++){
    newx[i] = 0;
    for(j=1;j<=n;j++){
        newx[i]=newx[i] + A[i][j] * x[j];
    }
    for(j=1;j<=m;j++){
        newx[i]=newx[i]+B[i][j]*u[j];
    }
}
for(i=1;i<=n;i++){
    x[i]=newx[i];
}
```

There are a few details to point out. We have gone to the trouble to calculate the new values of the variables as newx[] before updating the state. That is because each state must be calculated from the values the states arrived with. If x[1] has been changed before it is used in calculating x[2], we will get the wrong answer.

So why did we not go to this trouble in the continuous case? Because the increments were small and an approximation still holds good for small dt.

We could write the software more concisely but more cryptically by using += instead of repeating a variable on the right-hand side of an assignment. We could also use arrays that start from [0] instead of [1]. But the aim is to make the code as easy to read as possible.

Written in matrix form, all the problems appear incredibly easy. That is how it should be. A little more effort is needed to set up discrete equations corresponding to a continuous system, but even that is quite straightforward. Let us consider an example. Try it first, then read the next section to see it worked through.

Q 8.7.1

The output position x of a servomotor is related to the input u by a continuous equation:

$$\ddot{x} + \dot{x} = au,$$

i.e., when driven from one volt the output accelerates to a speed of a radians per second, with a settling time of one second. Express this in state variable terms.

It is proposed to apply computer control by sampling the error at intervals of 0.1 second, and applying a proportional corrective drive to the motor. Choose a suitable gain and discuss the response to an initial error. The system is illustrated in Figure 8.1.

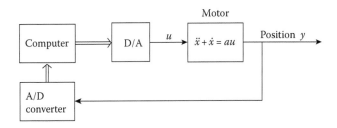

Figure 8.1 Position control by computer.

8.8 A Practical Example of Discrete Time Control

Let us work through example Q 8.7.1. The most obvious state variables of the motor are position x_1 and velocity x_2. The proposed feedback control uses position alone; we will assume that the velocity is not available as an output. We have

$$\dot{x}_1 = x_2$$
$$\dot{x}_2 = -x_2 + au$$

(8.4)

while

$$y = x_1.$$

We could attack the time-solution by finding eigenvectors and diagonalizing the matrix, then taking the exponential of the matrix and transforming it back, but that is really overkill in this case. We can apply "knife-and-fork" methods to see that if u is constant between samples, the velocity is of the form:

$$x_2 = be^{-t} + c$$

where c is proportional to u.

Substituting into the differential equation and matching initial conditions gives

$$x_2(t + \tau) = x_2(t)e^{-\tau} + a(1 - e^{-\tau})u.$$

By integrating $x_2(t)$ we find:

$$x_1(t + \tau) = x_1(t) + (1 - e^{-\tau})x_2(t) + a(\tau - 1 + e^{-\tau})u.$$

If we take τ to be 0.1 seconds, we can give numerical values to the coefficients. Since $e^{-0.1} = 0.905$, when we write $(n + 1)$ for "next sample" instead of $(t + \tau)$, we find that the equations have become

$$\begin{bmatrix} x_1(n+1) \\ x_2(n+1) \end{bmatrix} = \begin{bmatrix} 1 & 0.095 \\ 0 & 0.905 \end{bmatrix} \begin{bmatrix} x_1(n) \\ x_2(n) \end{bmatrix} + \begin{bmatrix} 0.005a \\ 0.095a \end{bmatrix} u(n).$$

Now we propose to apply discrete feedback around the loop, making $u(n) = -k\,x_1(n)$, and so we find that

$$\begin{bmatrix} x_1(n+1) \\ x_2(n+1) \end{bmatrix} = \begin{bmatrix} 1 & 0.095 \\ 0 & 0.905 \end{bmatrix}\begin{bmatrix} x_1(n) \\ x_2(n) \end{bmatrix} + \begin{bmatrix} 0.005a \\ 0.095a \end{bmatrix}\begin{bmatrix} -k & 0 \end{bmatrix}\begin{bmatrix} x_1(n) \\ x_2(n) \end{bmatrix}$$

that is,

$$\begin{bmatrix} x_1(n+1) \\ x_2(n+1) \end{bmatrix} = \begin{bmatrix} 1-0.005ak & 0.095 \\ -0.095ak & 0.905 \end{bmatrix}\begin{bmatrix} x_1(n) \\ x_2(n) \end{bmatrix}.$$

Next we must examine the eigenvalues, and choose a suitable value for k. The determinant gives us

$$\det\begin{vmatrix} 1-0.005ak-\lambda & 0.095 \\ -0.095ak & 0.905-\lambda \end{vmatrix} = 0,$$

so we have

$$\lambda^2 + (0.005ak-1.905)\lambda + (0.905+0.0045ak) = 0.$$

Well if $k = 0$, the roots are clearly 1 and 0.905, the open loop values.

If $ak = 20$ we are close to making the constant term exceed unity. This coefficient is equal to the product of the roots, so it must be less than one if both roots are to be less than one for stability. Just where within this range should we place k?

Suppose we look for equal roots, then we have the condition:

$$(0.005ak-1.905)^2 = 4(0.905+0.0045ak)$$

or multiplying through by 200^2,

$$(ak-381)^2 = 800(181+0.9ak)$$

$$(ak)^2 - (762+720)ak + (145161-144800) = 0,$$

i.e.,

$$(ak)2 - 1482ak + 361 = 0$$

$$ak = 741 \pm \sqrt{(741^2 - 361)} = 741 \pm 740.756.$$

Since ak must be less than 20, we must take the smaller value, giving a value of $ak = 0.244$—very much less than 20!

The roots will now be at $(1.905 - 0.005ak)/2$, giving a value of 0.952. This indicates that after two seconds, that is 20 samples, an initial position error will have decayed to $(0.952)^{20} = 0.374$ of its original value.

Try working through the next exercise before reading on to see its solution.

Q 8.8.1

What happens if we sample and correct the system of Q 8.7.1 less frequently, say at one second intervals? Find the new discrete matrix state equations, the feedback gain for equal eigenvalues and the response after two seconds.

Now the discrete equations have already been found to be

$$\begin{bmatrix} x_1(n+1) \\ x_2(n+1) \end{bmatrix} = \begin{bmatrix} 1 & 1-e^{-\tau} \\ 0 & e^{-\tau} \end{bmatrix} \begin{bmatrix} x_1(n) \\ x_2(n) \end{bmatrix} + \begin{bmatrix} a(\tau + e^{-\tau} - 1) \\ a(1 - e^{-\tau}) \end{bmatrix} u(n).$$

With τ set to unity, these become:

$$\begin{bmatrix} x_1(n+1) \\ x_2(n+1) \end{bmatrix} = \begin{bmatrix} 1 & 0.63212 \\ 0 & 0.36788 \end{bmatrix} \begin{bmatrix} x_1(n) \\ x_2(n) \end{bmatrix} + \begin{bmatrix} 0.36788a \\ 0.63212a \end{bmatrix} u(n),$$

then with feedback $u = -kx_1$,

$$\begin{bmatrix} x_1(n+1) \\ x_2(n+1) \end{bmatrix} = \begin{bmatrix} 1-0.36788ak & 0.63212 \\ -0.63212ak & 0.36788 \end{bmatrix} \begin{bmatrix} x_1(n) \\ x_2(n) \end{bmatrix}$$

yielding a characteristic equation:

$$\lambda^2 + (0.36788ak - 1.36788)\lambda + (0.26424ak + 0.36788) = 0$$

The limit of *ak* for stability has now reduced below 2.4 (otherwise the product of the roots is greater than unity), and for equal roots we have

$$(0.36788ak - 1.36788)^2 = 4(0.26424ak + 0.36788)$$

Pounding a pocket calculator gives *ak* = 0.196174—smaller than ever! With this value substituted, the eigenvalues both become 0.647856. It seems an improvement on the previous eigenvalue, but remember that this time it applies to one second of settling. In two seconds, the error is reduced to 0.4197 of its initial value, only a little worse than the previous control when the corrections were made 10 times per second.

Can we really get away with correcting this system as infrequently as once per second? If the system performs exactly as its equations predict, then it appears possible. In practice, however, there is always uncertainty in any system. A position servo is concerned with achieving the commanded position, but it must also maintain that position in the presence of unknown disturbing forces that can arrive at any time. A once-per-second correction to the flight control surfaces of an aircraft might be adequate to control straight-and-level flight in calm air, but in turbulence this could be disastrous.

8.9 And There's More

In the last chapter, when we considered feedback around an undamped motor we found that we could add stabilizing dynamics to the controller in the form of a *phase advance*. In the continuous case, this could include an electronic circuit to "differentiate" the position (but including a lag to avoid an infinite gain). In the case of discrete time control, the dynamics can be generated simply by one or two lines of software.

Let us assume that the system has an output that is simply the position, *x*. To achieve a damped response, we also need to feed back the velocity, *v*. Since we cannot measure it directly, we will have to estimate it. Suppose that we have already modeled it as a variable, *vest*.

We can integrate it to obtain a modeled position that we will call *xslow*:

```
xslow = xslow + vest*dt
```

So where did *vest* come from?

```
vest = k*(x-xslow)
```

Just these two lines of code are enough to estimate a velocity signal that can damp the system. Firstly, we will assume that *dt* is small and that this is an approximation of a continuous system.

$$\dot{x}_{slow} = v_{est}$$

and

$$v_{est} = k(x - x_{slow})$$

so

$$\dot{x}_{slow} = kx - kx_{slow}.$$

So in transfer function terms we have

$$x_{slow} = \frac{k}{s+k} x.$$

The transfer function represents a lag, so x_{slow} is well named. We find that

$$v_{est} = \frac{ks}{s+k} x$$

which is a version of the velocity lagged with time-constant $1/k$. When we use it as feedback, we might make

$$u = -fx - dv_{est}$$

so

$$u = \left(-f - \frac{dks}{s+k}\right)x,$$

i.e.,

$$u = -\frac{(f+dk)s + f}{s+k} x$$

and if we write $j\omega$ for s, we see the same sort of transfer function as the phase advance circuit at the end of Section 7.10.

Q 8.9.1

Test your answer to Q 7.10.1 by simulating the system in the Jollies framework.

Since our motor is a simple two-integrator system with no damping, our state variables x and v will be simulated by

```
x = x + x*dt
```

and

```
v = v + u*dt
```

We have already seen that we can estimate the velocity with

```
vest = k * (x-xslow)
xslow = xslow + vest * dt
```

and then we can set

```
u =-f*x-d*vest
```

Put these last three lines before the plant equations, since those need a value of u to have been prepared. You now have the essential code to put in the "step model" window.

Make sure that the declarations at the top of the code include all the variables you need, such as *f, d, vest,* and *xslow.* Then you can put lines of code in the initialize window to adjust all the parameters, together with something to set the initial conditions, and you have everything for a complete simulation.

When *dt* is larger, we must use a z-transform to obtain a more accurate analysis. But more of that later. For now, let us see how state space and matrices can help us.

8.10 Controllers with Added Dynamics

When we add dynamics to the controller it becomes a system in its own right, with state variables, such as x_{slow} and state equations of its own. These variables can be added to the list of the plant's variables and a single composite matrix equation can be formed.

Suppose that the states of the controller are represented by the vector **w**. In place of Equation 8.3.1, we will have

$$\mathbf{u}_n = \mathbf{F}\mathbf{y}_n + \mathbf{G}\mathbf{v}_n + \mathbf{H}\mathbf{w}_n,$$

while **w** can be the state of a system that has inputs of both **y** and **v**,

$$\mathbf{w}_{n+1} = \mathbf{K}\mathbf{w}_n + \mathbf{L}\mathbf{y}_n + \mathbf{P}\mathbf{v}_n$$

so when we combine these with the system equations

$$\mathbf{x}_{n+1} = \mathbf{M}\mathbf{x}_n + \mathbf{N}\mathbf{u}_n$$

and

$$\mathbf{y}_n = \mathbf{C}\mathbf{x}_n$$

we can make a composite matrix equation

$$\begin{bmatrix} \mathbf{x}_{n+1} \\ \mathbf{w}_{n+1} \end{bmatrix} = \begin{bmatrix} \mathbf{M} + \mathbf{NFC} & \mathbf{NG} \\ \mathbf{LC} & \mathbf{K} \end{bmatrix} \begin{bmatrix} \mathbf{x}_n \\ \mathbf{w}_n \end{bmatrix}$$

and here we are again with a matrix to be examined for eigenvalues, this time of an order higher than the original system.

Now we have built up enough theory to complete the design of the inverted pendulum experiment.

Chapter 9

Controlling an Inverted Pendulum

Although a subset of the apparatus can be made to perform as the position controller that we have already met, the problem of balancing a pendulum is fundamentally different. At least one commercial vendor of laboratory experiments has shown a total misunderstanding of the problem. We will analyze the system using the theory that we have met so far and devise software for computer control. We will also construct a simulation that shows all the important properties of the real system.

The literature is full of simulated solutions to artificial problems, but in the case of this pendulum, the same algorithms have been used by generations of students to control a real experiment that you could easily construct for your own laboratory.

So what should be the basis of such a strategy? Should we apply position control to the trolley that forms the base of the pendulum, with a demand defined by the pendulum's tilt? We will soon see that such an approach is completely wrong.

9.1 Deriving the State Equations

The "trolley" is moved by a pulley and belt system, just as was shown in Figure 4.4. A pendulum is pivoted on the trolley and a transducer measures the angle of tilt (Figure 9.1).

The analysis can be made very much easier if we make a number of assumptions. The first is that the tilt of the pendulum is small, so that we can equate the sine of the tilt to the value of the tilt in radians.

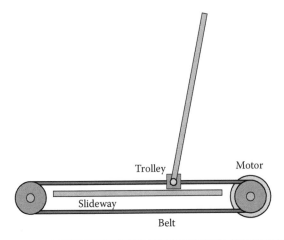

Figure 9.1   Inverted pendulum experiment.

The second is that the pendulum is much lighter than the trolley, so that we do not have to take into account the trolley's acceleration due to the tilt of the pendulum.

The third is that the motor is undamped, so that the acceleration is simply proportional to u, without any reduction as the speed increases. If the top speed of the motor is 6000 rpm, that is 100 revolutions per second, then without any gearing the top trolley speed could be over 10 meters per second—well outside any range that we will be working in.

These are reasonable assumptions, but the effects they ignore could easily be taken into account with a slight complication of the equations.

So what are our state variables?

For the trolley we have x and v, with the now familiar equations

$$\dot{x} = v$$

and

$$\dot{v} = bu$$

It is obvious that the tilt of the pendulum is a state variable, but it becomes clear when write an equation for its derivative that *tiltrate* must also be a variable—the rate at which the pendulum is falling over.

The tilt acceleration has a "falling over" term proportional to *tilt*, but how is it affected by the acceleration of the trolley?

Let us make another simplifying assumption, in which the pendulum consists of a mass on the top of a very light stick. (Otherwise we will get tangled up with moments of inertia and the center of gravity.)

The force in the stick is then mg, so the horizontal component will be $mg.\sin(tilt)$. The top will therefore accelerate at $g.\sin(tilt)$, or simply $g.tilt$, since the angle is small.

The acceleration of the tilt angle can be expressed as the difference between the acceleration of the top and the acceleration of the bottom, divided by the length L of the stick.

Since we know that the acceleration of the bottom of the stick is the acceleration of the trolley, or just bu, we have two more equations

$$\frac{d}{dt}\,tilt = tiltrate$$

and

$$\frac{d}{dt}\,tiltrate = (g.tilt - bu)/L.$$

In matrix form, this looks like

$$\frac{d}{dt}\begin{bmatrix} x \\ v \\ tilt \\ tiltrate \end{bmatrix} = \begin{bmatrix} 0 & 1 & 0 & 0 \\ 0 & 0 & 0 & 0 \\ 0 & 0 & 0 & 1 \\ 0 & 0 & \dfrac{g}{L} & 0 \end{bmatrix}\begin{bmatrix} x \\ v \\ tilt \\ tiltrate \end{bmatrix} + \begin{bmatrix} 0 \\ b \\ 0 \\ -\dfrac{b}{L} \end{bmatrix}u.$$

This looks quite easy so far, but we have to consider the effect of applying feedback. If we assume that all the variables have been measured, we can consider setting

$$u = c\,x + d\,v + e\,tilt + f\,tiltrate,$$

where c, d, e, and f are four constants that might be positive or negative until we learn more about them. When we substitute this value for u in the matrix state equation, it becomes

$$\frac{d}{dt}\begin{bmatrix} x \\ v \\ tilt \\ tiltrate \end{bmatrix} = \begin{bmatrix} 0 & 1 & 0 & 0 \\ bc & bd & be & bf \\ 0 & 0 & 0 & 1 \\ \dfrac{-bc}{L} & \dfrac{-bd}{L} & \dfrac{g-be}{L} & \dfrac{-bf}{L} \end{bmatrix}\begin{bmatrix} x \\ v \\ tilt \\ tiltrate \end{bmatrix}$$

and all we have to do is to find some eigenvalues.

We can save a lot of typing if we define the length of the stick to be one meter. To find the characteristic equation, we must then find the determinant and equate it to zero:

$$
\det \begin{vmatrix}
-\lambda & 1 & 0 & 0 \\
bc & bd - \lambda & be & bf \\
0 & 0 & -\lambda & 1 \\
-bc & -bd & g - be & -bf - \lambda
\end{vmatrix} = 0.
$$

Now we know that we can add one row to another without changing the value of a determinant, so we can simplify this by addition of the second row to the fourth:

$$
\det \begin{vmatrix}
-\lambda & 1 & 0 & 0 \\
bc & bd - \lambda & be & bf \\
0 & 0 & -\lambda & 1 \\
0 & -\lambda & g & -\lambda
\end{vmatrix} = 0.
$$

The task of expanding it now looks much easier. Expanding by the top row we have

$$
-\lambda \begin{vmatrix}
bd - \lambda & be & bf \\
0 & -\lambda & 1 \\
-\lambda & g & -\lambda
\end{vmatrix} - \begin{vmatrix}
bc & be & bf \\
0 & -\lambda & 1 \\
0 & g & -\lambda
\end{vmatrix} = 0.
$$

Q 9.1.1

Continue expanding the determinant to show that the characteristic equation is

$$
\lambda^4 + b(f - d)\lambda^3 + (b(e - c) - g)\lambda^2 + bgd\lambda + bgc = 0.
$$

Now, first and foremost we require all the roots to have negative real parts. This requires all the coefficients to be positive. The last two coefficients tell us that, perhaps surprisingly, the position and velocity feedback coefficients must be positive.

Since position control on its own would require large negative values, a controller based on position control of the trolley is going in entirely the wrong direction. It will soon become obvious why positive feedback is needed.

Now we see from the coefficient of λ^3 that the coefficient of *tiltrate* must exceed that of *v*. The coefficient of *tilt* must be greater than that of *x*, but must also overcome the "fall over" term. There are several other conditions, but we are still a long way from deciding actual values.

We can try the method of "pole assignment," choosing values for the roots and then matching corresponding coefficients of the characteristic polynomial. But where should we put the roots?

Let us try four roots of one second. This will give an equation:

$$(\lambda - 1)(\lambda - 1)(\lambda - 1)(\lambda - 1) = 0 \text{ or}$$

$\lambda^4 + 4\lambda^3 + 6\lambda^2 + 4\lambda + 1 = 0.$ So equating coefficients we have

$$b(f - d) = 4,$$

$$b(e - c) - g = 6,$$

$$bgd = 4,$$

$$bgc = 1.$$

We can approximate g to 10 and might also expect the trolley's acceleration to be 10 meters/second^2, so substituting these numbers will give

$$c = 0.01,$$

$$d = 0.04,$$

$$e = 1.61,$$

$$f = 0.44.$$

Now we can try this set of values in a simulation.

9.2 Simulating the Pendulum

We have already simulated a position control system that moves images on the screen. We can extend this to include the pendulum, as long as we just move the bob at the top—we cannot tilt the image of a rod.

First we calculate a value for u, so the code in the "step model" window will be

```
u = c*x + d*v + e*tilt + f*tiltrate;
```

Then we update the state variables with

```
x = x + v*dt;
v = v + b*u*dt;
tilt = tilt + tiltrate*dt;
tiltrate = tiltrate + g*tilt-b*u;
```

and finally move the picture elements to show the result:

```
Move(trolley, x, 0);
Move(bob, x + tilt, 1);
```

But we can make falling over look more realistic if instead of the last line we use

```
Move(bob, x + Math.sin(tilt), Math.cos(tilt));
```

For that matter, although we had to linearize the equations in order to look for eigenvalues we can simulate the system in all its non-linear glory by changing the *tiltrate* line to

```
tiltrate=tiltrate + g*Math.sin(tilt)-b*u*Math.cos(tilt);
```

so the whole of the code becomes

```
<html><head>
<title>Pendulum simulation</title>
<script language="javaScript">

function Move(a,b,c) {
   a.style.left=b*300+400;
   a.style.top=350-300*c;
}

var x=0.0;
var y=0.0;
var command=0.0;
var v=0.0;
var tilt=0.01;
var tiltrate=0;
var u=0.0;
var running=false;
var dt=.01;
var c, d, e, f;

function stepmodel(){
   eval(document.panel.model.value);}
</script>
</head>
```

```
<body style="background-color: rgb(255, 255, 204);">
<center>
<h2> Pendulum simulation </h2>
<br><br><br><br><br><br><br><br>
<br><br><br><br><br>
<form name="panel">

<input type="button" name="runit" value=" Run "
onclick="running=true;
         eval(document.panel.initial.value);
         stepmodel();">

<input type="button" name="stoppit" value=" Stop "
onclick="running=false;">

<b> Command</b>

<input type="button" value="-1" onclick=" command=-1;">
<input type="button" value=" 0 " onclick=" command=0;">
<input type="button" value=" 1 " onclick=" command=1;">
<br><br>
<textarea name="initial" rows="12" cols="60">
c = .01
d = .04
e = 1.61
f = .44
</textarea>

<textarea name="model" rows="12" cols="60">
//stepmodel()
u=c*(x-command)+d*v+e*tilt+f*tiltrate;

x=x+v*dt;
v=v+10*u*dt;
tilt=tilt+tiltrate*dt
tiltrate=tiltrate + (10*Math.sin(tilt)-10*u*Math.
cos(tilt))*dt;

Move(trolley, x, 0);
Move(bob, x + Math.sin(tilt), Math.cos(tilt));

//round again
if (running){setTimeout("stepmodel()",10);}
</textarea>
<br>

</form>
</center>

<div id="track"
style="position: absolute; left: 50; top: 410;">
<img src="road.gif" height="6" width="800"></div>
```

```
<div id="trolley"
style="position: absolute; left: 400; top: 350;">
<img src="block.gif" height="62" width="62"></div>

<div id="bob"
style="position: absolute; left: 400; top: 50;">
<img src="rball.gif" height="62" width="62"></div>

<script>
var trolley=document.getElementById("trolley");
var bob=document.getElementById("bob");
</script>
</body></html>
```

To go with the code, we have three images. Block.gif is the image of a cube that represents the trolley. Bob.gif is a picture of a ball, the pendulum "bob," while road. gif is a simple thick horizontal line for the slideway.

You will have noticed that the feedback gains have already been written into the "initial" text area. So how well does our feedback work?

If you do not want to go to the trouble of typing in the code, you can find everything at www.esscont.com/9/pend1.htm.

When you press "run" you see the trolley move slightly to correct the initial tilt. Perhaps you noticed that we gave *tilt* an initial value of 0.01. If we had started everything at zero, nothing would seem to happen when run is pressed, because the pendulum would be perfectly balanced in a way that is only possible in a simulation.

Pressing the "1" or "−1" buttons will command the trolley to move to the appropriate position, but of course it must keep the pendulum upright as it goes.

It appears to run just as expected, although with time constants of one second it seems rather sluggish.

9.3 Adding Reality

There is a slogan that I recite all too often,

"If you design a controller for a simulated system, it will almost certainly work exactly as expected—even though the control strategy might be disastrous on the real system."

So what is wrong with the simulation?

Any motor will have some amount of friction, even more so when it must drive a trolley along a track. It is an incredibly good motor and slide system for which this is only 5% of the motor full drive. We must therefore find a way to build friction into the simulation. It is best if we express the acceleration as a separate variable. We must also impose a limit of ± 1 on u.

Following the "u=" line, the code is replaced by

```
if (u>1){u=1;}
if (u<-1){u=-1;}
var accel=10*u;
if (v>0){accel=accel-.5;}
if (v<0){accel=accel+.5;}
x=x+v*dt;
v=v+accel*dt;
tilt=tilt+tiltrate*dt
tiltrate+=(10*Math.sin(tilt)-accel*Math.cos(tilt))*dt;
```

(That last line is written with the cryptic "+=" format shared by C and JavaScript to squeeze it onto one printed line.)

Change the code and test it, otherwise find the web version at www.esscont.com/9/pend2.htm.

The trolley dashes off the screen, to reappear hurrying in the opposite direction some seconds later. There is something wrong with our choice of poles.

9.4 A Better Choice of Poles

We can try to tighten up the response. Let us try two roots of one second, representing a sedate return of the trolley to a central position, and two more of 0.1 second to pull the pendulum swiftly upright. This will give an equation

$$(\lambda - 1)(\lambda - 1)(\lambda - 10)(\lambda - 10) = 0$$

or

$$\lambda^4 + 22\lambda^3 + 141\lambda^2 + 220\lambda + 100 = 0.$$

So equating coefficients we have

$$b(f - d) = 22,$$

$$b(e - c) - g = 141,$$

$$bgd = 220,$$

$$bgc = 100.$$

We can again approximate g to 10 and set the trolley's acceleration to be 10 meters/second^2, so substituting these new numbers will give

$$c = 1,$$

$$d = 2.2,$$

$$e = 16.1,$$

$$f = 4.4.$$

Try these new numbers in the simulation—the easy way is to change the values in the "initial" window. (Or load www.esscont.com/9/pend2a.htm)

The response is now much swifter and if the command is returned to zero the pendulum appears to settle. But after some 20 seconds the trolley will march a little to the side and later back again.

The friction of the motor has caused the response to display a limit cycle. In the case of position control, that same friction would enhance the system's stability with the penalty of a slight error in the settled position, but here the friction is destabilizing.

We can give the friction a more realistic value by changing the code to

```
if (v>0) {accel=accel-2;}
if (v<0) {accel=accel+2;}
```

and now the limit cycle is larger.

9.5 Increasing the Realism

We have drive limitation and friction, but the real system will suffer from further afflictions. Almost certainly the tilt sensor will have some datum error. Even if very small, it is enough to cause the system to settle with an error from center.

We can add a line

```
tiltsens = tilt + .001;
```

and use it instead of tilt when calculating feedback.

The next touch of realism is to recognize that although we can easily measure the tilt angle, for instance with a linear Hall effect device, we are going to have to deduce the rate of change of tilt instead of measuring it directly.

In Section 8.9 we saw that a rate signal could be generated just by a couple of lines of software,

```
xslow = xslow + vest*dt;
```

and

```
vest = k*(x-xslow);
```

We can use just the same technique to estimate *tiltrate*, using a new variable *trate* to hold the estimate

```
tiltslow = tiltslow + trate*dt
trate = k*(tiltsens-tiltslow)
```

but we now have to decide on a value for k. Recall that the smoothing time constant is of the order of $1/k$ seconds, but that $k\,dt$ should be substantially smaller than one. Here $dt = 0.01$ second, so we can first try $k = 20$.

Modify the start of the step routine to

```
tiltsens = tilt + .001;
tiltslow = tiltslow + trate*dt
trate = 20*(tiltsens-tiltslow)

u=c*(x-command)+d*v+e*tiltsens+f*trate;
```

Add

```
var trate=0;
var tiltslow=0;
var tiltsens=0;
```

to the declarations and try both the previous sets of feedback values.

(The code is saved at www.esscont.com/9/pend3.htm)

With the larger values, the motion of the trolley is "wobbly" as it oscillates around the friction, but the pendulum is successfully kept upright and the position moves as commanded.

Increasing the constant in the calculation of *trate* to 40 gives a great improvement. The extra smoothing time constant is significant if it is longer. It might be prudent to include it in the discrete state equations and assign all five of the resulting poles!

What is clear is that the feedback parameters for four equal roots are totally inadequate when the non-linearities are taken into account.

We can add a further embellishment by assuming that there is no tacho to measure velocity. Now we can add the lines

```
xslow = xslow + vest*dt;
```

and

```
vest = 10*(x-xslow);
```

to run our system on just the two sensors of tilt and position. Once again we have to declare the new variables and change to the line that calculates *u* to

```
u=c*(x-command)+d*vest+e*tiltsens+f*trate;
```

This is saved at www.esscont.com/9/pend4.htm

Is pole assignment the most effective way to determine the feedback parameters, or is there a more pragmatic way to tune them?

9.6 Tuning the Feedback Pragmatically

Let us design the feedback step by step. What is the variable that requires the closer attention, the trolley position, or the tilt of the pendulum? Clearly it is the latter. Without swift action the pendulum will topple. But if we feed tilt and tilt-rate back to the motor drive, with positive coefficients, we can obtain a rapid movement to bring the pendulum upright. But we will need an experiment to test the response.

In practice we would grip the top of the pendulum and move it to and fro, seeing the trolley move beneath it like "power assisted steering."

In the simulation we can use the command input in a similar way. If we force a proportion 0.2 of the command value onto the horizontal position of the bob, then the tilt angle will be (0.2*command-x), assuming small angles and a length of one meter. Even if we calculate *tiltrate* as before, it should play no part in changing the tilt angle. However, it could integrate up to give an error, so we reset it to zero for the response test.

It is the calculated *trate* that we feed back to the motor that matters now.

So at the top of the window add

```
tilt=0.2*command-x;
tiltrate=0;
```

and remove *command* from the line that calculates *u,*

```
u=c*x+d*vest+e*tiltsens+f*trate;
```

We are only interested in the tilt response, so in the initialize window we can set

```
c=0;
d=0;
```

and experiment with values of *e* and *f* to get a "good" response. (Remember that any changes will only take effect when you "stop" and "run" again.)

Experiments show that *e* and *f* can be increased greatly, so that values of

```
e=100;
f=10;
```

or even

```
e=1000;
f=100;
```

return the pendulum smartly upright.

So what happens when we release the bob?

To do that, simply "remark out" the first two lines by putting a pair of slashes in front of them

```
//tilt=0.2*command-x;
//tiltrate=0;
```

Now we see that the pendulum is kept upright, but the trolley accelerates steadily across the screen. What causes this acceleration?

Our tilt sensor has a deliberately introduced error. If this were zero, the pendulum could remain upright in the center and the task would seem to have been achieved. The offset improves the reality of the simulation.

But now we need to add some strategy to bring the trolley to the center. Pragmatically, how should we think of the appropriate feedback?

If the pendulum tilts, the trolley accelerates in that direction. To move the trolley to the center, we should therefore cause the pendulum to "lean inwards." For a given position of the pendulum bob, we would require the trolley to be displaced "outwards." That confirms our need for positive position feedback.

"Grip" the bob again by removing the slashes from the two lines of code. Try various values of c, so that the pendulum tilts inwards about a "virtual pivot" some three meters above the track. Try a value for c of 30.

Release the bob again. Now the pendulum swings about the "virtual pivot," but with an ever-increasing amplitude. We need some damping.

We find that with these values of e and f, c must be reduced to 10 to achieve reasonable damping for any value of d. But when we increase e and f to the improbably high values that we found earlier, a good response can be seen for

```
c = 50
d = 100
e = 1000
f = 100
```

These large values, coupled with the drive constraint, mean that the control will be virtually "bang-bang," taking extreme values most of the time. An immediate consequence is that the motor friction will be of much less importance.

The calibration error in the tilt sensor, will however lead to a settling error of position that is 20 times as great as the error in radians. The "virtual pivot" for this set of coefficients is some 20 meters above the belt.

9.7 Constrained Demand

The state equations for the pendulum are very similar to those for the exercise of riding a bicycle to follow a line. The handlebars give an input that accelerates the rear wheel perpendicular to the line, while the "tilt" equations for falling over are virtually the same.

When riding the bicycle, we have an "inner loop" that turns the handlebars in the direction we are leaning, in order to remain upright. We then relate any desire to turn in terms of a corresponding demand for "lean" in that same direction. But self-preservation compels us to impose a limit on that lean angle, otherwise we fall over.

We might try that same technique here.

Our position error is now (*command-x*), so we can demand an angle of tilt

```
tiltdem=c*(command-x)-d*vest;
```

which we then constrain with

```
if(tiltdem>.1){tiltdem=.1;}
if(tiltdem<-.1){tiltdem=-.1;}
```

and we use tiltdem to compute *u* as

```
u=e*(tilt-tiltdem)+f*trate;
```

Note that our values of *c* and *d* will not be as they were before, since *tiltdem* is multiplied by feedback parameter *e*. For a three-meter height of the virtual pivot, *c* will be 1/3.

First try $c=0.3$ and $d=0.2$. Although the behavior is not exactly stable, it has a significant feature. Although the position of the trolley dithers around, the tilt of the pendulum is constrained so that there is no risk of toppling.

In response to a large position error the pendulum is tilted to the limit, causing the trolley to accelerate. As the target is neared, the tilt is reversed in an attempt to bring the trolley to a halt. But if the step change of position is large, the speed builds up so that overshoots are unavoidable.

So that suggests the use of another limited demand signal, for the trolley velocity.

```
vdem=c*(command-x)
if(vdem>vlim){vdem=vlim;}
if(vdem<-vlim){vdem=-vlim;}
```

and a new expression for tiltdem

```
tiltdem=d*(vdem-vest);
```

We add definitions

```
var vlim=.5;
var vdem=0;
```

plus a line in the initialize box to change *vlim*'s value.
Try the values

```
c = 2;
d = .1;
e = 1000;
f = 100;
vlim=1;
```

When the trolley target is changed, there is a "twitch" of the tilt to its limit that starts the trolley moving. It then moves at constant speed across the belt toward the target, then settles after another twitch of tilt to remove the speed.

You can find the code at www.esscont.com/9/pend6.htm

9.8 In Conclusion

Linear mathematical analysis may look impressive, but when the task is to control a real system some non-linear pragmatism can be much more effective. The methods can be tried on a simulation that includes the representation of drive-limitation, friction, discrete time effects, and anything else we can think of. But it is the phenomena that we do not think of that will catch us out in the end. Unless a strategy has been tested on a real physical system, it should always be taken with a pinch of salt.

ESSENTIALS OF CONTROL THEORY— WHAT YOU OUGHT TO KNOW

II

Chapter 10

More Frequency Domain Background Theory

So far we have seen how to analyze a real system by identifying state variables, setting up state equations that can include constraints, friction, and all the aspects that set practice apart from theory and write simulation software to try out a variety of control algorithms. We have seen how to apply theory to linear systems to devise feedback that will give chosen response functions. We have met complex exponentials as a shortcut to unscrambling sinusoids and we have seen how to analyze and control systems in terms of discrete time signals and transforms.

This could be a good place to end a course in practical control.

It is a good place to start an advanced course in the theory that will increase a fundamental mathematical understanding of methods that will impress.

10.1 Introduction

In Chapter 7, complex frequencies and corresponding complex gain functions were appearing thick and fast. The gains were dressed up as transfer functions, and poles and zeros shouted for attention. The time has come to put the mathematics onto a sound footing.

By considering complex gain functions as "mappings" from complex frequencies, we can exploit some powerful mathematical properties. By the application of Fourier and Laplace transforms to the signals of our system, we can avoid taking unwarranted liberties with the system transfer function, and can find a slick way to deal with initial conditions too.

10.2 Complex Planes and Mappings

A complex number $x + jy$ is most easily visualized as the point with coordinates (x, y) in an "Argand diagram." In other words, it is a simple point in a plane. The points on the x-axis, where $y = 0$, are real. The points on the y-axis, where $x = 0$, are pure imaginary. The rest are a complex mixture.

If we represent $x + jy$ by the single symbol z, then we can start to consider functions of z that will also be complex numbers. For a start, consider

$$w = z^2. \qquad\qquad (10.1)$$

Here, w is another complex number that can be represented as a point in a plane of its own. For any point in the z-plane, there is a corresponding point in the w-plane. Things get more interesting when we consider the mapping not just of a single point, but of a complete line.

If z is a positive imaginary number, ja, then in our example w will be $-a^2$. Any point on the positive imaginary axis of the z-plane will map to a point on the negative real axis of the w-plane. As a varies from zero to infinity, w moves from zero to minus infinity, covering the entire negative axis. We see that any point on the negative imaginary z-axis will also map to the negative real w-plane axis. (A mathematician would say that the function mapped the imaginary z-axis "onto" the negative real axis, but "into" the entire real axis.)

Points on the z-plane real axis will map to the w-plane positive real axis whether z is positive or negative. To find points which map to the w-plane imaginary axis, we have to look toward values of z of the form $a(1 + j)$ or $a(1 - j)$. Other lines in the z-plane map to curves in the w-plane.

Q 10.2.1

Show that as a varies, $z = (1 + ja)$ maps to a parabola.

We need to find out more about the way such mappings will distort more general shapes. If we move z slightly, w will be moved too. How are these displacements related? If we write the new value of w as $w + \delta w$, then in the example of Q 10.2.1, we have

$$w + \delta w = (z + \delta z)^2,$$

i.e.,

$$\delta w = 2z\delta z + \delta z^2.$$

If δz is small, the second term can be ignored. Around any given value of z, δw is the product of δz, and a complex number—in this case $2z$. More generally, we

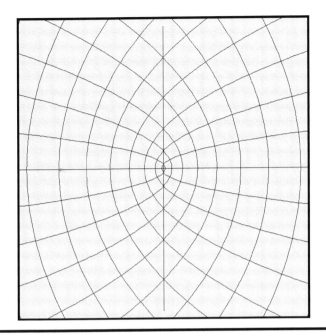

Figure 10.1   Illustration of mapping defined in Q 10.2.1.

can find a local derivative *dw/dz* by the same algebra of small increments that is used in any introduction to differentiation.

Now, when we multiply two complex numbers we multiply their amplitudes and add their "arguments." Multiplying any complex number by $(1 + j)$, for example, will increase its amplitude by a factor of $\sqrt{2}$, while rotating its vector anticlockwise by 45°. Take four small δz's, which take *z* around a small square in the *z*-plane, and we see that *w* will move around another square in the *w*-plane, with a size and orientation dictated by the local derivative *dw/dz*. If *z* moves around its square in a clockwise direction, *w* will also move round in clockwise direction.

Of course, as larger squares are considered, the *w*-shapes will start to curve. If we rule up the *z*-plane as squared graph-paper, the image in the *w*-plane will be made up of "curly squares" as seen in Figure 10.1.

The illustration is a screen-grab from the code at www.esscont.com/10/z2mapping.htm

10.3 The Cauchy–Riemann Equations

In the previous section, we suggested that a function of *z* could be differentiated in the same way that a real derivative could be found. This is true as far as "ordinary" functions are concerned, ones that would be found in "real calculus." However, there are many more functions that can be applied to a complex variable which do not make sense for a real one.

Consider the function $w = \text{Real}(z)$, just the real part, x. Now w will always lie on the real axis, and any thought of it moving in squares will make no sense. This particular function and many more like it can possess no derivative.

If a function $w(z)$ is to have a derivative at any given point, then for a perturbation δz in z, we must be able to find a complex number $(A + jB)$ so that around that point $\delta w = (A + jB)\delta z$. Let us consider the implications of this requirement, first looking at our example in Q 10.2.1.

w can be expressed as $u + jv$, so we can expand the expression to show

$$u + jv = (x + jy)^2$$

$$= x^2 - y^2 + 2jxy.$$

Separating real and imaginary parts, we see the real part of w,

$$u = x^2 - y^2,$$

and the imaginary part

$$v = 2xy. \tag{10.2}$$

In general, u and v will be functions of x and y, a relationship that can be expressed as:

$$u = u(x, y)$$
$$v = v(x, y)$$

If we make small changes δx and δy in the values of x and y, the resulting changes in u and v will be defined by the *partial derivatives* in the equations:

$$\delta u = \frac{\partial u}{\partial x}\delta x + \frac{\partial u}{\partial y}\delta y$$

and

$$\delta v = \frac{\partial v}{\partial x}\delta x + \frac{\partial v}{\partial y}\delta y. \tag{10.3}$$

By holding the change in y to zero, we can clearly see that $\partial u/\partial x$ is the gradient of u, if x is changed alone. Similarly, the three other partial derivatives are gradients when just one variable is changed.

Now,

$$\delta w = \delta u + j\delta v$$

$$= \left(\frac{\partial u}{\partial x}\delta x + \frac{\partial u}{\partial y}\delta y \right) + j\left(\frac{\partial v}{\partial x}\delta x + \frac{\partial v}{\partial y}\delta y \right)$$

$$= \left(\frac{\partial u}{\partial x} + j\frac{\partial v}{\partial x} \right)\delta x + \left(\frac{\partial u}{\partial y} + j\frac{\partial v}{\partial y} \right)\delta y \qquad (10.4)$$

$$= \left(\frac{\partial u}{\partial x} + j\frac{\partial v}{\partial x} \right)\delta x + \left(\frac{\partial v}{\partial y} - j\frac{\partial u}{\partial y} \right)j\delta y.$$

If w is to have a derivative, we must be able to write

$$\delta w = (A + jB)(\delta x + j\delta y),$$

so we can equate coefficients to obtain

$$A = \frac{\partial u}{\partial x} = \frac{\partial v}{\partial y}$$

and

$$B = \frac{\partial v}{\partial x} = -\frac{\partial u}{\partial y}. \qquad (10.5)$$

Leave out A and B and these are the Cauchy–Riemann equations.

The main interest of such mappings from the control point of view is the "curly squares" approximation for estimating fine detail. The Cauchy–Riemann equations partly express the condition for a function to be "analytic." Beware, though. Analytic behavior may be restricted to portions of the plane, and strange things happen at a singularity such as a pole.

Q 10.3.1

Show that the Cauchy–Riemann equations hold for the functions of the example worked out in Equations 10.2.

10.4 Complex Integration

In real calculus, integration can be regarded as the inverse of differentiation. If we want to evaluate the integral of $f(x)$ from 1 to 5, we look for a function $F(x)$ of which $f(x)$ is the derivative. We work out $F(5) - F(1)$, and are satisfied to write that down as the answer. Will the same technique work for complex integration?

In the real case, this integral is the limit of the sum of small contributions

$$f(x)\delta x,$$

where the total of the δx values takes x from 1 to 5. There is clearly just one answer (for a "well behaved" function). In the complex case, we can define a similar integral as the sum of contributions

$$f(z)\delta z.$$

This time the answer is not so cut and dried. The trail of infinitesimal δz's must take us from $z = 1$ to $z = 5$, but now we are not constrained to the real axis but can wander around the z-plane. We have to define the path, or "contour," along which we intend to integrate, as seen in Figure 10.2.

With luck, Cauchy's theorem can come to our rescue. If we can find a simple curve that encloses a region of the z-plane in which the function is everywhere *regular* (analytic without singularities), then it can be shown that any path between the endpoints lying completely within the region will give the same answer. This

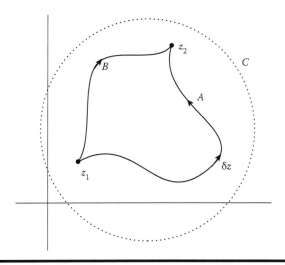

Figure 10.2 **Contour integrals in the *z*-plane.**

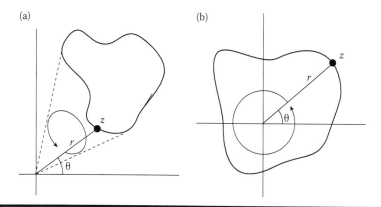

Figure 10.3 Evaluating log(z) around a contour.

is the same as saying that the integral around a loop in the region, starting, and ending at the same point, will be zero.

If our path encircles a pole, it is a different matter entirely. Consider the case where $f(z)$ is $1/z$ and our path is an anticlockwise circle around the origin from $z = 1$ to $z = 1$ again. We know that the integral of $1/z$ is $\ln(z)$ (the "natural" log to base e of z) so we can write down the answer as $\ln(1) - \ln(1)$. But this will not be zero!

The logarithm is not a single-valued function when we are dealing with complex variables. If we express z as $re^{j\theta}$ then we see that $\ln(z) = \ln(r) + j\theta$. If we take an anticlockwise trip around the origin, the vector representing z will have rotated through a complete circle and the value of θ will have increased by 2π. Our integral would therefore have the value $2\pi j$.

We could take another trip around the circle to increase θ to $4\pi j$, and so on. With anticlockwise or clockwise circuits, θ can take any value $2n\pi j$, where n is an integer positive or negative. Clearly θ will clock up a change of $2\pi j$ if z takes any trip around the origin, the singularity, but for a journey around a small local circuit $\theta\kappa$ will return to the same value, as illustrated in Figure 10.3.

In the case of $f(z) = 1/z$, the integral in question will have the value $2\pi j$. For $f(z) = 7/z$, the integral would be $14\pi j$. For $f(z) = \cos(z)/z$, the integral would have the value $2\pi j$—since $\cos(0) = 1$. In general, the integral around a simple pole at $z = a$ will be $2\pi j$ times the *residue* at the pole, that is the limiting value of $(z - a)$ $f(z)$ as z tends to a.

The residue can be thought of as the value of the function when the offending factor of the denominator is left out. We can turn this around the other way and say that if $f(z)$ is regular inside the unit circle (or any other loop for that matter), then we can find out the value of $f(a)$ (a is inside the loop) by integrating $f(z)/(z - a)$ around the loop to obtain $2\pi j f(a)$. It seems to be doing things the hard way, but it has its uses.

Before we leave this topic, let us establish some facts for a later chapter. Integrating $1/z$ around the unit circle gives the value $2\pi j$. Integrating $1/z^2$ gives the value zero. Its general integral is $-1/z$, and this is no longer multi-valued. In fact, the integral of z^n around the circle for any integer n will be zero, except for $n = -1$.

If $f(z)$ can be represented by a power series in $1/z$,

$$f(z) = \sum a_n z^{-n},$$

then we can pick out the coefficient an by multiplying $f(z)$ by z^{n-1} and integrating around the unit circle. This gives us a transform between sequences of values an and a function of z. We can construct the function of z simply by multiplying the coefficients by the appropriate powers of z and summing (all right, there is the question of convergence), and we can unravel the coefficients again by the integral method.

This transform between a sequence of values and a function of z will become important when we consider sampled data systems in more detail. But first there are two more transforms that are relevant to continuous systems, the Laplace transform and the Fourier transform.

10.5 Differential Equations and the Laplace Transform

In Chapter 2 we glanced at the problems of finding analytic solutions to differential equations. Now we will look at an alternative approach that will give a rigorous solution.

Suppose we have a function of time, $f(t)$, and that we are really only interested in its values for $t > 0$. For reasons we will see later, we can multiply $f(t)$ by the exponential function e^{-st} and integrate from $t = 0$ to infinity. We can write the result as

$$F(s) = \int_0^\infty f(t)e^{-st}\,dt.$$

The result is written as $F(s)$, because since the integral has been evaluated over a defined range of t, t has vanished from the resulting expression. On the other hand, the result will depend on the value chosen for s. In the same way that we have been thinking of $f(t)$ not as just one value but as a graph of $f(t)$ plotted against time, so we can consider $F(s)$ as a function defined over the entire range of s. We have transformed the function of time, $f(t)$, into a function of the variable s. This is the "unilateral" Laplace transform.

Consider an example. The Laplace transform of the function $5e^{-at}$ is given by

$$\int\limits_{0}^{\infty} 5e^{-at}e^{-st}dt$$

$$= \int\limits_{0}^{\infty} 5e^{-(s+a)t}dt$$

$$= \frac{-5}{s+a}\left[e^{-(s+a)t}\right]_{0}^{\infty}$$

$$= \frac{5}{s+a}.$$

We have a transform that turns functions of t into functions of s. Can we work the trick in reverse? Given a function of s, can we find just one function of time of which it is the transform? We may or may not arrive at a precise mathematical process for finding the "inverse"—it is sufficient to spot a suitable function, provided that we can show that the answer is unique.

For "well-behaved" functions, we can show that the transform of the sum of two functions is the sum of the two transforms:

$$\mathcal{L}\{f(t)+g(t)\} = \int\limits_{0}^{\infty}\{f(t)+g(t)\}e^{-st}dt$$

$$= \int\limits_{0}^{\infty}\{f(t)\}e^{-st}dt + \int\limits_{0}^{\infty}\{g(t)\}e^{-st}dt$$

$$= F(s)+G(s).$$

Now suppose that two functions of time, $f(t)$ and $g(t)$, have the same Laplace transform. Then the Laplace transform of their difference must be zero:

$$\mathcal{L}\{f(t)-g(t)\} = F(s)-G(s) = 0$$

since we have assumed $F(s) = G(s)$. What values can $[f(t) - g(t)]$ take, if its Laplace integral is zero for every value of s? It can be shown that if we require $f(t) - g(t)$ to be differentiable, then it must be zero for all $t > 0$; in other words the inverse transform (if it exists) is unique.

Why should we be interested in leaving the safety of the time domain for these strange functions of s? Consider the transform of the derivative of $f(t)$:

$$\mathcal{L}\{f'(t)\} = \int\limits_{0}^{\infty} f'(t)e^{-st}dt.$$

Integrating by parts, we see

$$L\{f'(t)\} = [f(t)e^{-st}]_0^\infty - \int_0^\infty f(t)\frac{d}{dt}(e^{-st})dt$$

$$= -f(0) - (-s)\int_0^\infty f(t)e^{-st}dt$$

$$= sF(s) - f(0).$$

We can use this result to show that

$$L\{f''(t)\} = sL\{f'(t)\} - f'(0)$$

$$= s^2 F(s) - sf'(0) - f'(0)$$

and in general

$$L\{f^{(n)}(t)\} = s^n F(s) - s^{n-1} f(0) - s^{n-2} f'(0) \dots f^{(n-1)}(0).$$

So what is the relevance of all this to the solution of differential equations? Suppose we are faced with

$$\ddot{x} + x = 5e^{-at} \qquad (10.6)$$

Now, if $L\{x(t)\}$ is written as $X(s)$, we have

$$L\{\ddot{x}\} = s^2 X(s) - sx(0) - \dot{x}(0)$$

so we can express the transform of the left-hand side as

$$L\{\ddot{x} + x\} = (s^2 + 1)X(s) - sx(0) - \dot{x}(0).$$

For the right-hand side, we have already worked out that

$$L\{5e^{-at}\} = \frac{5}{s+a},$$

so

$$(s^2 + 1)X(s) - sx(0) - \dot{x}(0) = \frac{5}{s+a},$$

or

$$X(s) = \frac{1}{s^2 + 1}\left[\frac{5}{s+a} + sx(0) + \dot{x}(0)\right]. \tag{10.7}$$

Without too much trouble we have obtained the Laplace transform of the solution, complete with initial conditions. But how do we unravel this to give a time function? Must we perform some infinite contour integration or other? Not a bit!

The art of the Laplace transform is to divide the solution into recognizable fragments. They are recognizable because we can match them against a table of transforms representing solutions to "classic" differential equations. Some of the transforms might have been obtained by infinite integration, as we showed with e^{-at}, but others follow more easily by looking at differential equations.

The general solution to

$$\ddot{x} + x = 0$$

is

$$x = A\cos(t) + B\sin(t).$$

Now

$$\mathcal{L}\{\ddot{x} + x\} = 0$$

so

$$X(s) = \frac{1}{s^2 + 1}[sx(0) + \dot{x}(0)].$$

For the function $x = \cos(t)$, $x(0) = 1$, and $\dot{x}(0) = 0$, then

$$\mathcal{L}\{\cos(t)\} = \frac{s}{s^2 + 1}.$$

If $x = \sin(t)$, $x(0) = 0$, and $\dot{x}(0) = 1$, then

$$\mathcal{L}\{\sin(t)\} = \frac{1}{s^2 + 1}.$$

With these functions in our table, we can settle two of the terms of Equation 10.7. We are left, however, with the term:

$$\frac{1}{s^2 + 1} \cdot \frac{5}{s+a}.$$

Using partial fractions, we can crack it apart into

$$\frac{A + Bs}{s^2 + 1} + \frac{C}{s + a}.$$

Before we know it, we find ourselves having to solve simultaneous equations to work out A, B, and C. These are equivalent to the equations we would have to solve for the initial conditions in the straightforward "particular integral and complementary function" method.

Q 10.5.1

Find the time solution of differential equation 10.6 by solving for A, B, and C in transform 10.7 and substituting back from the known transforms. Then solve the original differential equation the "classic" way and compare the algebra involved.

The Laplace transform really is not a magical method of solving differential equations. It is a systematic method of reducing the equations, complete with initial conditions, to a standard form. This allows the solution to be pieced together from a table of previously recognized functions. Do not expect it to perform miracles, but do not underestimate its value.

10.6 The Fourier Transform

After battling with Laplace, the Fourier transform may seem rather tame. The Laplace transformation involved the multiplication of a function of time by e^{-st} and its integration over all positive time. The Fourier transform requires the time function to be multiplied by $e^{-j\omega t}$ and then integrated over all time, past, and future.

It can be regarded as the analysis of the time function into all its frequency components, which are then presented as a frequency spectrum. This spectrum also contains phase information, so that by adding all the sinusoidal contributions the original function of time can be reconstructed.

Start by considering the Fourier series. This can represent a repetitive function as the sum of sine and cosine waves. If we have a waveform that repeats after time $2T$, it can be broken down into the sum of sinusoids of period $2T$, together with their harmonics.

Let us set out by constructing a repetitive function of time in this way. Rather than fight it out with sines and cosines, we can allow the function to be complex, and we can take the sum of complex exponentials:

$$f(t) = \sum_{n=-\infty}^{\infty} c_n e^{n\frac{\pi}{T}jt}. \tag{10.8}$$

Can we break $f(t)$ back down into its components? Can we evaluate the coefficients cn from the time function itself?

The first thing to notice is that because of its cyclic nature, the integral of $e^{n\pi jt/T}$ from $t = -T$ to $+T$ will be zero for any integer n except zero. If $n = 0$, the exponential degenerates into a constant value of unity and the value of the integral will be just $2T$.

Now if we multiply $f(t)$ by $e^{-r\pi t/T}$, we will have the sum of terms:

$$f(t)e^{-r\pi t/T} = \sum_{n=-\infty}^{\infty} c_n e^{(n-r)\pi t/T}.$$

If we integrate over the range $t = -T$ to $+T$, the contribution of every term on the right will vanish, except one. This will be the term where $n = r$, and its contribution will be $2Tc_r$.

We have a route from the coefficients to the time function and a return trip back to the coefficients.

We considered the Fourier series as a representation of a function which repeated over a period $2T$, i.e., where its behavior over $t = -T$ to $+T$ was repeated again and again outside those limits. If we have a function that is not really repetitive, we can still match its behavior in the range $t = -T$ to $+T$ with a Fourier series (Figure 10.4).

The lowest frequency in the series will be π/T. The series function will match $f(t)$ over the center range, but will be different outside it. If we want to extend the range of the matching section, all we have to do is to increase the value of T. Suppose that we double it, then we will halve the lowest frequency present, and the interval between the frequency contributions will also be halved. In effect, the number of contributing frequencies in any range will be doubled.

We can double T again and again, and increase the range of the match indefinitely. As we do so, the frequencies will bunch closer and closer until the summation of Equation 10.8 would be better written as an integral. But an integral with respect to what?

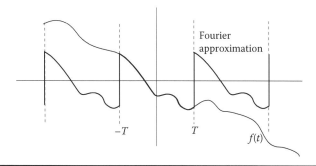

Figure 10.4 A Fourier series represents a repetitive waveform.

We are interested in frequency, which we write in angular terms as ω. The frequency represented by the nth term of Equation 10.8 is $n\pi/T$, so let us write this as ω and let us write the single increment π/T as $\delta\omega$. Instead of the set of discrete coefficients c_n we will consider a continuous function $F(j\omega)$ of frequency.

As we let T tend to infinity we arrive at

$$F(j\omega) = \int_{-\infty}^{\infty} f(t)e^{-j\omega t} dt \qquad (10.9)$$

and

$$f(t) = \frac{1}{2\pi} \int_{-\infty}^{\infty} F(j\omega)e^{j\omega t} d\omega. \qquad (10.10)$$

We have a clearly defined way of transforming to the frequency domain and back to the time domain. You may have noticed a close similarity between the integral of Equation 10.9 and the integral defining the Laplace transform. Substitute s for $j\omega$ and they are closer still. They can be thought of as two variations of the same integral, where in the one case s takes only real values, while in the other its values are purely imaginary.

It will be of little surprise to find that the two transforms of a given time function appear algebraically very similar. Substitute $j\omega$ for s in the Laplace transform and you usually have the Fourier transform. They still have their separate uses. This introduction sets the scene for the transforms. There is much more to learn about them, but that comes later.

Q 10.6.1

Find the Fourier transform of $f(t)$, where

$$f(t) = \begin{cases} e^{-at} & \text{if} \quad t > 0 \\ 0 & \text{if} \quad t < 0 \end{cases}.$$

Chapter 11

More Frequency Domain Methods

11.1 Introduction

In Chapter 7, we saw the engineer testing a system to find how much feedback could be applied around it before instability set in. We saw that simple amplitude or power measurements enabled a frequency response to be plotted on log–log paper, leading to the development of rules of thumb to become analytic methods backed by theory. We saw too that the output amplitude told only half the story, and that it was important also to measure the phase.

In the early days, phase measurement was not easy. The input and output waveforms could be compared on an oscilloscope, but the estimate of phase was somewhat rough and ready. If an $x - y$ oscilloscope was used, output could be plotted as y against the input's x, enabling phase shifts of zero and multiples of 90° to be accurately spotted. The task of taking a complete frequency response was a tedious business (Figure 11.1).

Then came the introduction of the phase-sensitive voltmeter, often given the grand title of *Transfer Function Analyzer*. This contained the sine-wave source to excite the system, and bore two large meters marked *Reference* and *Quadrature*. By using a *synchronous demodulator*, the box analyzed the return signal into its components in-phase and 90° out-of-phase with the input. These are the real and imaginary parts of the complex amplitude, properly signed positive or negative.

It suddenly became easy to measure accurate values of complex gain, and the Nyquist diagram was straightforward to plot. With the choice of Nyquist, Bode, Nichols, and Whiteley, control engineers could argue the benefits of their particular

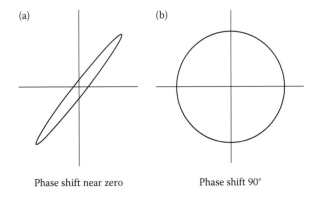

(a)

(b)

Phase shift near zero

Phase shift 90°

Figure 11.1   Oscilloscope measurement of phase.

favorite. Soon time-domain and *pseudo-random binary sequence* (PRBS) test methods were adding to the confusion—but they have no place in this chapter.

11.2 The Nyquist Plot

Before looking at the variety of plots available, let us remind ourselves of the object of the exercise. We have a system that we believe will benefit from the application of feedback. Before "closing the loop," we cautiously measure its open loop frequency response (or transfer function) to ensure that the closed loop will be stable. As a bonus, we would like to be able to predict the closed loop frequency response.

Now if the open loop transfer function is $G(s)$, the closed loop function will be

$$\frac{G(s)}{1+G(s)} \tag{11.1}$$

This is deduced as follows. If the input is $U(s)$ and we subtract the output $Y(s)$ from it in the form of feedback, then the input to the "inner system" is $U(s) - Y(s)$. So

$$Y(S) = G(S)\{U(S) - Y(S)\}$$

i.e.,

$$\{1 + G(s)\}Y(s) = G(s)U(s)$$

so

$$Y(s) = \frac{G(s)}{1+G(s)}U(s)$$

We saw that stability was a question of the location of the poles of a system, with disaster if any pole strayed to the right half of the complex frequency plane. Where will we find the poles of the closed loop system? Clearly they will lie at the values of s that give $G(s)$ the value -1. The complex gain $(-1 + j0)$ is going to become the focus of our attention.

If we plot the readings from the phase-sensitive voltmeter, the imaginary part against the real with no reference to frequency, we have a Nyquist plot. It is the path traced out in the complex gain plane as the variable s takes value $j\omega$, as ω increases from zero to infinity. It is the image in the complex gain plane of the positive part of the imaginary s axis.

$$\text{Suppose that } G(s) = \frac{1}{1 + s}$$

then

$$
\begin{aligned}
G(j\omega) &= \frac{1}{1 + j\omega} \\
&= \frac{1 - j\omega}{1 + \omega^2} \\
&= \frac{1}{1 + \omega^2} - j\frac{\omega}{1 + \omega^2}
\end{aligned}
$$

If G is plotted in the complex plane as $u + jv$, then it is not hard to show that

$$u^2 + v^2 - u = 0$$

This represents a circle, though for "genuine" frequencies with positive values of ω we can only plot the lower semicircle, as shown in Figure 11.2. The upper half of the circle is given by considering negative values of ω. It has a diameter formed by the line joining the origin to $(1 + j0)$. What does it tell us about stability?

Clearly the gain drops to zero by the time the phase shift has reached $90°$, and there is no possible approach to the critical gain value of -1. Let us consider something more ambitious.

The system with transfer function

$$G(s) = \frac{1}{s(s + 1)(s + 1)} \tag{11.2}$$

has a phase shift that is never less than $90°$ and approaches $270°$ at high frequencies, so it could have a genuine stability problem. We can substitute $s = j\omega$ and manipulate the expression to separate real and imaginary parts:

$$G(j\omega) = \frac{1}{j\omega(1 - \omega^2 + 2 j\omega)}$$

So multiplying the top and bottom by the conjugate of the denominator, to make the denominator real, we have

$$G(j\omega) = \frac{-2\omega^2 - j\omega(1 - \omega^2)}{\omega^2(1 + \omega^2)^2}$$

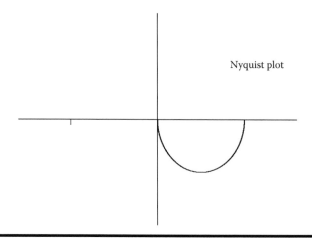

Figure 11.2   Nyquist plot of 1/(1 + s). (Screen grab from www.esscont.com/11/ nyquist.htm)

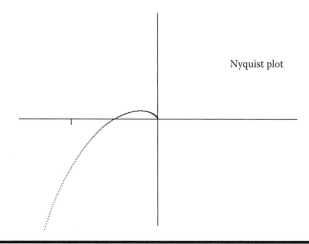

Figure 11.3   Nyquist plot of 1/s(s + 1)². (Screen grab from www.esscont.com/11/ nyquist2.htm)

The imaginary part becomes zero at the value $\omega = 1$, leaving a real part with value −1/2. Once again, algebra tells us that there is no problem of instability. Suppose that we do not know the system in algebraic terms, but must base our judgment on the results of measuring the frequency response of an unknown system. The Nyquist diagram is shown in Figure 11.3. Just how much can we deduce from it?

Since it crosses the negative real axis at −0.5, we know that we have a gain margin of 2. We can measure the phase margin by looking at the point where it crosses the unit circle, where the magnitude of the gain is unity.

11.3 Nyquist with M-Circles

We might wish to know the maximum gain we may expect in the closed loop system. As we increase the gain, the frequency response is likely to show a "resonance" that will increase as we near instability. We can use the technique of M-circles to predict the value, as follows.

For unity feedback, the closed loop output $Y(j\omega)$ is related to the open loop gain $G(j\omega)$ by the relationship

$$Y = \frac{G}{1+G} \qquad (11.3)$$

Now Y is an analytic function of the complex variable G, and the relationship supports all the honest-to-goodness properties of a mapping. We can take an interest in the circles around the origin that represent various magnitudes of the closed loop output, Y. We can investigate the G-plane to find out which curves map into those constant-magnitude output circles.

We can rearrange Equation 11.3 to get

$$Y + YG = G$$

so

$$G = \frac{Y}{1-Y}$$

By letting Y lie on a circle of radius m,

$$Y = m(\cos\theta + j\sin\theta)$$

we discover the answer to be another family of circles, the M-circles, as shown in Figure 11.4. This can be shown algebraically or simply by letting the software do the work; see www.esscont.com/mcircle.htm.

Q 11.3.1

By letting $G = x + jy$, calculating Y, and equating the square of its modulus to m, use Equation 11.3 to derive the equation of the locus of the M-circles in G.

We see that we have a safely stable system, although the gain peaks at a value of 2.88.

We might be tempted to try a little more gain in the loop. We know that doubling the gain would put the curve right through the critical −1 point, so some smaller value must be used. Suppose we increase the gain by 50%, giving an open loop gain function:

$$G(s) = \frac{1.5}{s(s+1)^2}$$

In Figure 11.5 we see that the Nyquist plot now sails rather closer to the critical −1 point, crossing the axis at −0.75, and the M-circles show there is a resonance with closed loop gain of 7.4.

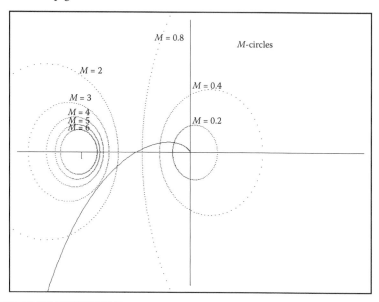

Figure 11.4   M-circles.

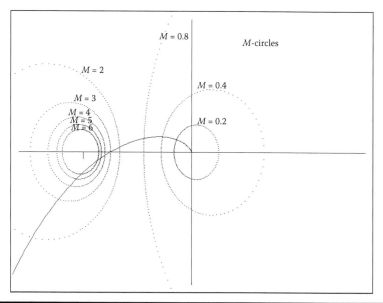

Figure 11.5   Nyquist with higher gain and M-circles. (Screen-grab from www. esscont.com/11/mcircle2.htm)

11.4 Software for Computing the Diagrams

We have tidied up our software by making a separate file, *jollies.js*, of the routines for the plotting applet. We can make a further file that defines some useful functions to handle the complex routines for calculating a complex gain from a complex frequency. We can express complex numbers very simply as two-component arrays.

The contents of *complex.js* are as follows. First we define some variables to use. Then we define functions to calculate the complex gain, using functions to copy, subtract, multiply, and divide complex numbers. We also have a complex log function for future plots.

```
var gain=[0,0];          // complex gain
var denom=[1,0];         // complex denominator for divide
var npoles;              // number of poles
var poles = new Array(); // complex values for poles
var nzeros;              // number of zeroes
var zeros = new Array(); // complex values for zeroes
var s=[0,0];             // complex frequency, s
var vs=[0,0];            // vector s minus pole or s minus zero
var temp=0;
var k=1;                 // Gain multiplier for transfer
                         // function

function getgain(s){     // returns complex gain for complex s
  gain= [k,0];           // Initialise numerator
  for(i=0;i<nzeros; i++){
    csub(vs,s,zeros[i]); // ds = s-zero[]
    cmul(gain,vs);       // multiply vectors for numerator
  }
  denom= [1,0];          // initialise denominator
  for(i=0;i<npoles; i++){
    csub(vs,s,poles[i]);
    cmul(denom,vs);      // multiply vectors from poles
  }
  cdiv(gain,denom);      // divide numerator by denominator
}
function copy(a,b)       {// complex b = a
  b[0]=a[0];
  b[1]=a[1];
}
function csub(a,b,c){     // a = b-c
  a[0]=b[0]-c[0];
  a[1]=b[1]-c[1];
}
function cmul(a,b){       // a = a * b
  temp=a[0]*b[0]-a[1]*b[1];
```

```
   a[1]=a[0]*b[1]+a[1]*b[0];
   a[0]=temp;
}
function cdiv(a,b){        // a = a / b
   temp=a[0]*b[0]+a[1]*b[1];
   a[1]=a[1]*b[0]-a[0]*b[1];
   a[0]=temp;
   temp=b[0]*b[0]+b[1]*b[1];
   a[0]=a[0]/temp;
   a[1]=a[1]/temp;
}
function clog(a){          // a = complex log(a)
   temp=a[0]*a[0]+a[1]*a[1];
   temp=Math.log(temp)/2;
   a[1]=Math.atan2(a[1],a[0]);
   a[0]=temp;
}
```

The code does not require very much editing to change it to give the other examples.

11.5 The "Curly Squares" Plot

Can we use the open loop frequency response to deduce anything about the closed loop time-response to a disturbance? Surprisingly, we can.

Remember that the plot shows the mapping into the G-plane of the $j\omega$ axis of the s-plane. It is just one curve in the mesh that would have to be drawn to represent the "mapped graph-paper" effect of Figure 10.1. Remember also that the squares of the coordinate grid of the s-plane must map into "curly squares" in the G-plane.

Let us now tick off marks along the Nyquist curve to represent equal increments in frequency, say of 0.1 radians per second, and build onto these segments a mosaic of near-squares. We will have an approximation to the mapping not only of the imaginary axis, but also of a "ladder" formed by the imaginary axis, by the vertical line $-0.1 + j\omega$, and with horizontal "rungs" joining them at intervals of $0.1j$, as shown in Figure 11.6.

Now we saw in Section 7.7 that the response to a disturbance will be made up of terms of the form $\exp(pt)$, where p is a pole of the overall transfer function—and in this case we are interested in the closed loop response. The closed loop gain becomes infinite only when $G=-1$, and so any value of s which maps into $G=-1$ will be a pole of the closed loop system.

Looking closely at the "curved ladder" of our embroidered Nyquist plot, we see that the -1 point lies just below the "rung" of $\omega=0.9$, and just past half way across it. We can estimate reasonably accurately that the image of the -1 point in the s-plane

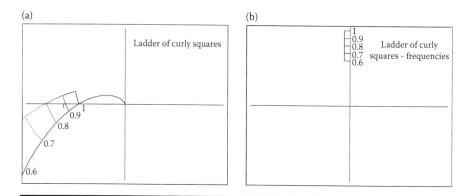

Figure 11.6   Curly squares and rungs for 1.5/(s(s+1)²). (Screen grabs from www. esscont.com/11/ladderfreq.htm and www.esscont.com/11/ladder.htm)

has coordinates $-0.055+j\,0.89$. We therefore deduce that a disturbance will result in a damped oscillation with frequency 0.89 radians per second and decay time constant $1/0.055=18$ seconds.

(If you are worried that poles should come in complex conjugate pairs, note that the partner of the pole we have found here will lie in the corresponding image of the negative imaginary s axis, the mirror image of this Nyquist plot, which is usually omitted for simplicity.)

Q 11.5.1

By drawing "curly squares" on the plot of $G(s)=1/(s(s+1)^2)$ (Figure 11.3), estimate the resonant frequency and damping factor for unity feedback when the multiplying gain is just one. Note that the plot will be just 2/3 of the size of the one in Figure 11.6.

(Algebra gives $s=-0.122+j\,0.745$. You could simply look at Figure 11.6 and see where the point $(-1.5,0)$ falls in the curly mesh!).

Q 11.5.2

Modify the code of Nyquist2.htm to produce the curly squares plot. Look at the source of ladder.htm to check your answer. Why has omega/10 been used?

11.6 Completing the Mapping

With mappings in mind, we can be a little more specific about the conditions for stability. We can regard the plot not just as the mapping of the positive imaginary s axis, but of a journey outward along the imaginary axis. As s moves upward in the example of Figure 11.3, G move from values in the lower left quadrant in toward the G-plane origin. As it passes the -1 point, it lies on the left-hand side of the path,

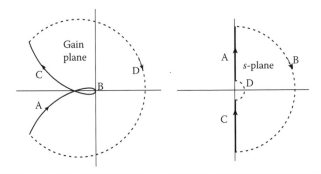

Figure 11.7 Nyquist plot for $1/(s(1+s)^2)$ "completed" with negative frequencies.

implying from the theory of complex functions that s leaves the corresponding pole on the left-hand side of the imaginary axis—the safe side.

We can extend this concept by considering a journey in the s-plane upwards along the imaginary axis to a very large value, then in a clockwise semicircle enclosing the "dangerous" positive half-plane, and then back up the negative imaginary axis to the origin. In making such a journey, the mapped gain must not encircle any poles if the system is to be stable. This results in the requirement that the "completed" G-curve must not encircle the −1 point.

If G becomes infinite at $s = 0$, as in our present example, we can bend the journey in the s-plane to make a small anticlockwise semicircular detour around $s=0$, as shown in Figure 11.7.

11.7 Nyquist Summary

We have seen a method of testing an unknown system, and plotting the in-phase and quadrature parts of the open loop gain to give an insight into closed loop behavior. We have not only a test for stability, by checking to see if the −1 point is passed on the wrong side, but an accurate way of measuring the peak gains of resonances. What is more, we can in many cases extend the plot by "curly squares" to obtain an estimate of the natural frequency and damping factor of a dangerous pole.

This is all performed in practice without a shred of algebra, simply by plotting the readings of an "$R \& Q$" meter on linear graph-paper, estimating closed loop gains with the aid of pre-printed M-circles.

11.8 The Nichols Chart

The $R \& Q$ meter lent itself naturally to the plotting of Nyquist diagrams, but suppose that the gain data was obtained in the more "traditional" form of gain and phase, as used in the Bode diagram. Would it be sensible to plot the logarithmic gain directly, gain against the phase-angle, and what could be the advantages?

We have seen that an analytic function has some useful mathematical properties. It can also be shown that an analytic function of an analytic function is itself analytic. Now the logarithm function is a good honest analytic function, where

$$\log(G(s)) = \log(|G|) + j \arg(G).$$

(Remember that the function "arg" represents the phase-angle in radians, with value whose tangent is Imag(G)/Real(G). The atan2 function, taking real and imaginary parts of G as its input parameters, puts the result into the correct quadrant of the complex plane.)

Instead of plotting the imaginary part of G against the real, as for Nyquist, we can plot the logarithmic gain in decibels against the phase shift of the system. All the rules about encircling the critical point where $G=-1$ will still hold, and we should be able to find the equivalents of M-circles.

The point $G=-1$ will of course now be defined by a phase shift of π radians or 180°, together with a gain of 0 dB. The M-circles are circles no longer, but since the curves can be pre-printed onto the chart paper, that is no great loss. The final effect is shown in Figure 11.8.

So what advantage could a Nichols plot have over a Nyquist diagram?

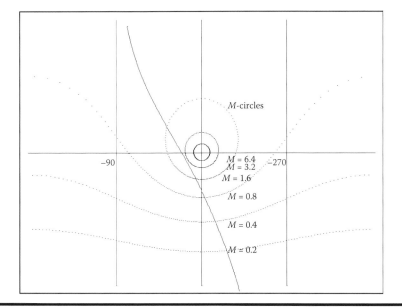

Figure 11.8 Nichols chart for 1/(s(s+1)²). (Screen grab of www.esscont.com/11/ nichols.htm)

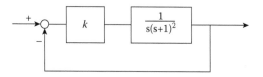

Figure 11.9   Block diagram with unity feedback around a variable gain.

Suppose that we wish to consider a variety of gains applied before the open loop transfer function input, as defined by k in Figure 11.9. To estimate the resulting closed loop response, we might have to rescale a Nyquist diagram for each value considered. Changing the gain in a Nichols plot, however, is simply a matter of moving the plot vertically with respect to the chart paper. By inspecting Figure 11.8 it is not hard to estimate the value of k, for instance, which will give a peak closed loop gain of 3.

When more sophisticated compensation is considered, such as phase-advance, the Nichols chart relates closely to the Bode diagram and the two can be used together to good effect.

11.9   The Inverse-Nyquist Diagram

There is an alternative to the Nyquist diagram that maintains simplicity while making it easy to relate open loop to closed loop gain. It is sometimes called the Whiteley diagram.

We have become accustomed to thinking in terms of gain, applying one volt to the input of a circuit and measuring the output. An equally valuable concept is inverse gain. What input voltage will give an output of just one volt? When we start closing loops, we see that inverse gain is much simpler to deal with. Consider a system with open loop gain G and unity feedback, as in Figure. 11.10.

If the output is one unit, then the input to the G-element is $1/G$. The feedback is again one unit, so the input to the closed loop system is $1 + 1/G$. If we use the symbol W for inverse gain, then

$$W_{\text{closed loop}} = W_{\text{open-loop}} + 1.$$

That's really all there is to it. The frequency response of the closed loop system is obtained from the open loop just by moving it one unit to the right. In the open loop plot, the -1 point is still the focus of attention. When the loop is closed, this moves to the origin. The origin represents one unit of output for zero input—just as we would expect for infinite gain.

The M-circles are still circles, but now they are simply centered on the -1 point. Larger circles imply lower gains. A radius of 2 implies a closed loop gain of $1/2$, and

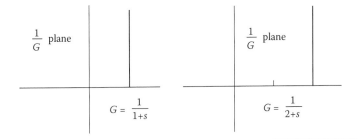

Figure 11.10 Signals in terms of an output of unity.

<table>
<tr><td>$\frac{1}{G}$ plane</td><td></td><td>$\frac{1}{G}$ plane</td></tr>
<tr><td>$G = \frac{1}{1+s}$</td><td></td><td>$G = \frac{1}{2+s}$</td></tr>
</table>

Figure 11.11 Whiteley plots for $G=1/(1+s)$ and $G=1/(2+s)$.

so on. At the same time, the phase shift is simply given by measuring the angle of the vector joining the point in question to the −1 point—and then negating the answer.

Some of the more familiar transfer functions become almost ridiculously easy. The lag

$$G(s) = \frac{1}{1+s}$$

which gave a semicircular Nyquist plot, now appears as

$$W(s) = 1+s$$

When we give s a range of values $j\omega$, the Whiteley plot is simply a line rising vertically from the point $W=1$. We close the loop with unity feedback and see that the line has moved one unit to the right to rise from $W=2$. (See Figure 11.11.)

With more complicated functions, we might become worried about the "rest" of the plot, for negative frequencies and for the large complex frequencies that complete the loop in the s-plane. With Nyquist, we usually have no need to bother, since the high-frequency gain generally drops to zero and the plot muddles gently around the origin, well away from the −1 point. The inverse gain often becomes infinite, however, and the plot may soar around the boundaries of the diagram.

In the $1/(1+s)$ example, $W(s)$ approximates to s for large values of s, and so as s makes a clockwise semicircular detour around the positive-real half-plane, so W will make a similar journey well away from the -1 point, as shown in Figure 11.12.

It is now clear that the inverse-Nyquist plot is at its best when we wish to consider a variable gain k in the feedback loop. (See Figure 11.13.) Since the closed loop gain is now the inverse of $1/G+k$, we can slide the Whiteley plot any distance to the right to examine any particular value of k.

The damped motor found in examples of position control has a transfer function of the form:

$$G(s) = \frac{b}{s(s+a)}$$

so the inverse gain is

$$W(j\omega) = j\omega(j\omega+a)/b$$

$$= -\omega^2/b + aj\omega/b.$$

The plot is a simple parabola. The portion of the plot for negative frequencies completes the symmetry of the parabola, leaving us only to worry about very large complex frequencies. Now for large s, W approximates to s^2 and so as s makes its

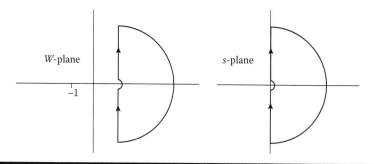

Figure 11.12   Whiteley plot for $G=1/(1+s)$ with loop around the positive s-plane.

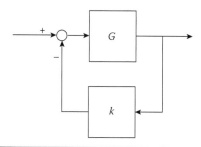

Figure 11.13   Block diagram with feedback gain of k.

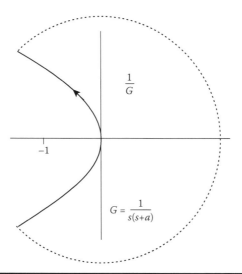

Figure 11.14   Whiteley plot of 1/s(s+a) with mapping of frequency plane "return semicircle" added.

return journey around a large semicircle of radius R, from $+jR$ clockwise via $+R$ to $-jR$, W will move from $-R^2$ clockwise through double the angle via $+R^2$ and round through negative imaginary values to $-R^2$ again. The entire plot is shown in Figure 11.14.

Clearly since it does not cut the real axis anywhere to the left of the -1 point, no amount of feedback will result in instability.

To obtain an illustration that could go unstable, we have been using the third-order system

$$G(s) = \frac{1}{s(s+1)^2}$$

What does this look like as an inverse-Nyquist diagram?

$$W(j\omega) = j\omega (j\omega + 1)^2$$

$$= -2\omega^2 + j\omega(1 - \omega^2).$$

As ω increases from zero, the plot starts from the origin. The real part starts from zero and becomes more negative. The imaginary part increases slightly, then crosses the axis when $\omega = 1$ and the real part $= -2$. From there, both real and imaginary parts become increasingly negative. The curve then dives off South–West on a steepening curve. When we plot the values for negative frequency, it looks suspiciously as

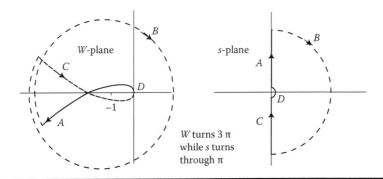

Figure 11.15 Inverse-Nyquist with path completed.

though the −1 point has been surrounded-disaster! But we must take great care to examine the return semicircular journey.

W approximates to the cube of *s* for large values, and so as *s* moves around its semicircular π radians, *W* will move in the same sense through 3π radians. When *s* is "at the top" of the imaginary axis, *W* has a large negative imaginary part and a somewhat smaller negative real part. As *s* rotates clockwise, so *W* rotates clockwise through one-and-a-half revolutions to join the negative-frequency plot high up and rather to the left of the imaginary axis, as shown in Figure 11.15. If the plot were a loop of cotton, and a pencil were stabbed into the −1 point, then the loop could be pulled clear without wrapping around the pencil. Despite first appearances the −1 point is not encircled, and the system is stable.

If we consider a feedback gain of 3, however, we would have to look at the point *W*=−3. This is well and truly encircled, twice in fact, and the system would be unstable.

11.10 Summary of Experimental Methods

It must be repeated that the methods of these last two chapters relate to experimental testing of an unknown system. Although most of the examples have been based on simple transfer functions, this is merely for convenience of explanation and under-standing. In practice, the engineer would have to deduce any transfer function from a jumbled table of readings and would have to make an intelligent decision about the feedback.

In the case of the Bode plot, an attempt is made to recognize "breakpoints" so that a guess can be made at the transfer function. Gain variations can be consid-ered by simply raising or lowering the log-gain plot, and rule-of-thumb will get the engineer quite a long way. Even without recognizing the transfer function, some quite sophisticated compensators can be prescribed, in the form of phase-advance or lag-lead filters. Although stability is well defined in terms of gain margin or phase margin, however, prediction of the performance of the closed loop system is not easy.

The Nyquist plot takes the output from an "R & Q" meter and displays it quite simply as imaginary part of gain against real. M-circles enable the peak closed loop gain to be accurately predicted. Gain values for marginal stability are easily deduced too, although deducing closed loop response for anything but unity feedback involves some work. There are rigorous rules to deduce the stability of complicated compensators that cause the gain to exceed unity at a phase shift of 180°, yet pull the phase back again to avoid encircling the −1 point.

The Nichols plot is most simply applied to readings in the form of log-gain and phase-angle. It offers all the benefits of Nyquist, barring the distortion of the M-curves, and allows easy consideration of variations in gain. When contemplating phase-advance and other compensators, it makes a good partner to a Bode plot.

The inverse-Nyquist plot will usually require some calculation in order to turn the readings into plotted points; when a computer is involved, this is no handicap. It offers the same tests and criteria as Nyquist, but gives a much clearer insight when considering variable feedback. Some head-scratching may be involved in deducing the "closing" plot for large complex frequencies.

Q 11.10.1

Draw an inverse-Nyquist diagram for the system:

$$G(s) = \frac{1}{s^2(s+1)}$$

Can it be stabilized by proportional feedback alone?

Q 11.10.2

In Section 5.6, a water-heater experiment was introduced. Its transfer function combines a lag with a time-delay. Both time constants are 10 seconds, and the gain is such that if continuous full power were to be commanded by setting $u = 1$, the water temperature would be raised by 100°. The transfer function is therefore

$$G(s) = \frac{10e^{-10s}}{s+0.1}.$$

Sketch an inverse-Nyquist plot and estimate the maximum loop gain that can be applied before oscillation occurs.

Whenever we have defined an example by its transfer function, there has been the temptation to bypass the graphics and manipulate the algebra to solve for resonant frequencies, limiting gains and so on. In the next chapter we will see how to take full advantage of such knowledge, and to look at the stability not just of one particular feedback gain, but of an entire range so that we can make an intelligent choice.

Chapter 12

The Root Locus

12.1 Introduction

In the last chapter, we saw that a reasonably accurate frequency response allowed us to plan the feedback gain. We could even make an educated guess at some of the poles of the closed loop system. If we have deduced or been given the transfer function in algebraic terms, we can use some more algebra to compute the transfer function of the closed loop system and to deduce the values of all its poles and zeros.

Faced with an intimidating list of complex numbers, however, we might still not find the choice of feedback gain to be very easy. What we need is some way of visualizing in graphic terms the effect of varying the feedback. What we need is the root locus plot, the possible locations of all the roots as we vary the feedback gain.

12.2 Root Locus and Mappings

When we examined an inverse-Nyquist diagram, we saw that a closed loop pole for unity feedback could be assessed from the location of the -1 point within the "curly squares" mapping of the frequency plane. In order to examine the effect of a feedback gain k, we could look at the point $-k + j0$ instead. For variable feedback gain, we can consider any point along this negative real axis. We could consider the positive real axis, too, but this would imply positive feedback.

Instead of pondering the "curly squares" which represent the mapping of the s-plane "graph-paper" onto the plane of complex inverse gain, can we map the real gain axis back into the frequency plane? That would give us all the possible combinations of the closed loop poles.

Consider the system

$$G(s) = \frac{1}{s(s+1)}$$

or in differential equation form

$$\ddot{x} + \dot{x} = u$$

This is a "damped motor" kind of system, with a rather slow time-constant of one second and an integrator that relates velocity to position.

It has one pole at the origin and another at $s = -1$.

If we apply feedback $u = -kx$ as shown in Figure 12.1, we will achieve some sort of position control, described by the roots of

$$s^2 + s + k = 0$$

As we increase the feedback from zero, the closed loop poles leave the open loop poles and move toward each other along the negative real axis, until they meet when $k = 1/4$. Here we have critical damping.

As we increase k even further, the poles split and move north and south, a complex pair with values

$$s = -\frac{1}{2} \pm j\sqrt{k - \frac{1}{4}}$$

The decay time-constant remains the same, but the frequency of the resonance increases as k gets bigger.

We will be able to see all this if we can map the negative real gain axis back into the complex frequency plane, and that is easy to do with our routines for calculating gains.

For points of s on a grid we can calculate the gain. If the imaginary part is positive, we display a square red "blob" at the coordinates of s, otherwise we show a yellow one. When we have painted the whole plane, we will have boundaries between the red and yellow areas that represent places where the imaginary part of the gain changes sign—a locus on which the gain is real.

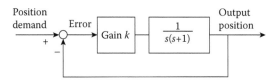

Figure 12.1 Position control with variable feedback.

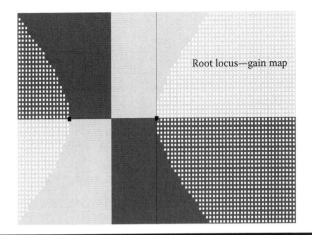

Root locus—gain map

Figure 12.2 Gain map for 1/s(s+1). (Screen-grab from www.esscont.com/12/ gainmap1.htm)

But we have not quite finished. The boundary will correspond to all values of k, both positive and negative. So we put a smaller blob, white this time, at each point where the real part of the gain is positive. When we have marked in the open loop poles, the locus of the roots for negative feedback is clearly seen in Figure 12.2.

The "code in the window" of the.htm file for creating this is

```
npoles=2;
poles[0]= [0,0];
poles[1]= [-1,0];
nzeros=0;

var ds=.04;
var dw=.04;
function Blob(x,c){
  Colour(c);
  BoxFill(x[0]-ds/2,x[1]+ds/2,x[0]+ds/2,x[1]-dw/2);
}
function Blip(x,c){
  Colour(c);
  BoxFill(x[0]-ds/4,x[1]+ds/4,x[0]+ds/4,x[1]-dw/4);
}
function Showroots(){
  for(i=0;i<npoles;i++){
    Blob(poles[i],Black);
  }
  for(i=0;i<nzeros;i++){
    Blob(zeros[i],Black);
    Blip(zeros[i],White);
```

```
    }
  }
InitGraph();
ScaleWindow(-1.6,-1,1.6,1);
DrawAxes();
Showroots();
for(s[1]=-.98;s[1]<1;s[1]+=dw){
  for(s[0]=-1.58;s[0]<1.6;s[0]+=ds){
    getgain(s);
    if(gain[1]>0){
      Blob(s,Red);
    }else{
      Blob(s,Yellow);
    }
    if(gain[0]>0){
      Blip(s,White);
    }
  }
}
DrawAxes();
Showroots();
Colour(Black);
Label('Root locus-gain map',.5,.5);
```

When we add a second pole at $-1+j0$, the map becomes somewhat more complicated.

Now in Figure 12.3 we can see the pair of poles at -1 split apart in opposite directions along the real axis. One moves off toward negative infinity, where it can do no harm. The other one moves to meet the pole from the origin at a point that

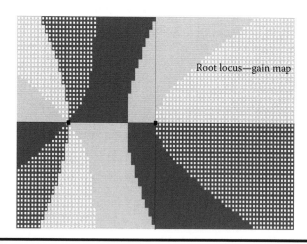

Root locus—gain map

Figure 12.3 Gain map of $1/s(s+1)^2$.

appears to be −1/3. From there they split upward and down, but instead of moving along paths parallel to the imaginary axis they curve to the right, passing into the right-hand half-plane. The value of k that gives equal roots would certainly appear to be the best choice.

These diagrams are striking, especially when seen in color, but they are an unusual way to portray the actual locus. Using our gain calculation routine and some insight into mappings, we can draw a root locus that looks more conventional.

12.3 A Root Locus Plot

The sledgehammer approach is to substitute an array of values of k into the equation, then use a polynomial root solver to give the points to plot. Instead we can refine the "gain map" technique to obtain the same effect much more simply.

In constructing the gain map, we have swept though an array of s-values in a "raster." The imaginary part of s defines a row of points in the s-plane. The real part of s then sweeps from left to right. For the gain map we just planted a blob, but instead we can detect a change of sign of the imaginary part of the gain to indicate when we have crossed the root locus. We can even compare the gain values either side of the change to interpolate and get a much more accurate value of s than its step change. We could then plant a dot. As we add increments to the imaginary part of s to move the row upward, we obtain a rather spotty locus.

Some of the boundaries might be parallel to the real axis, so we would also need to sweep a second raster in vertical lines across the plane to be sure of completing the picture.

But we can do better still. We are trying to map the real gain axis. A mapping back to the frequency plane will take each small segment of this axis and rotate it by the complex derivative *ds/dgain*. But the only function that we have is a mapping in the other direction, from s to *gain*. That does not matter, since *dgain/ds* is just the inverse of what we are looking for. As we step the real part of s along, the difference between the new and previous gains will give us just two of the partial derivatives we need. But the Cauchy–Riemann equations give us the rest!

With center at the *segmid* point that we have found for the spot plot, we can draw a small segment of length *ds*. To find the angle at which to plot the segment, we rotate it by the reverse of the angle found for *dgain*, the gain change between steps. It might be clearer to look at the code itself.

```
npoles=3;
poles[0]= [0,0];
poles[1]= [-1,0];
poles[2]= [-1,0];
nzeros=0;
```

```
var ds=.04;
var segmid=[0,0];
var dgain=[0,0];
var oldgain=[0,0];
var segment=[0,0];
function Blob(x,c){
  Colour(c);
  BoxFill(x[0]-ds/2,x[1]+ds/2,x[0]+ds/2,x[1]-ds/2);
}
function Blip(x,c){
  Colour(c);
  BoxFill(x[0]-ds/4,x[1]+ds/4,x[0]+ds/4,x[1]-ds/4);
}
function Showroots(){
  for(i=0;i<npoles;i++){
    Blob(poles[i],Black);
  }
for(i=0;i<nzeros;i++){
  Blob(zeros[i],Black);
  Blip(zeros[i],White);
  }
}
function getsegment(){
  mod=Math.sqrt(dgain[0]*dgain[0]+dgain[1]*dgain[1]);
  if(mod>0){
    segment[0]=ds*dgain[0]/mod;
    segment[1]=-ds*dgain[1]/mod;
  }else{
    segment=[0,0];
  }
}
InitGraph();
ScaleWindow(-1.6,-1,1.6,1);
DrawAxes();
Showroots();
Colour(Black);
for(s[1]=-.98;s[1]<1;s[1]+=ds){//raster one way
  segmid[1]=s[1];
  oldgain[1]=0;
  for(s[0]=-1.58;s[0]<1.6;s[0]+=ds){
    getgain(s);
    if((gain[1]*oldgain[1]<0)&&(gain[0]<0)){
      csub(dgain,gain,oldgain);
      getsegment();
      segmid[0]=s[0]-ds*gain[1]/dgain[1];
      LineStart(segmid[0]-segment[0]/2,segmid[1]-segment[1]/2);
      LineTo(segmid[0]+segment[0]/2,segmid[1]+segment[1]/2);
    }
```

```
    copy(gain,oldgain);
  }
}
for(s[0]=-1.58;s[0]<1.6;s[0]+=ds){//then the other way
  segmid[0]=s[0];
  oldgain[1]=0;
  for(s[1]=-.98;s[1]<1;s[1]+=ds){
    getgain(s);
    if((gain[1]*oldgain[1]<0)&&(gain[0]<0)){
      csub(dgain,gain,oldgain);
      getsegment();
      segmid[1]=s[1]-ds*gain[1]/dgain[1];
      LineStart(segmid[0]+segment[1]/2,segmid[1]-segment[0]/2);
      LineTo(segmid[0]-segment[1]/2,segmid[1]+segment[0]/2);
    }
    copy(gain,oldgain);
  }
}
Label("Root locus",.5,.5);
```

The functions for showing the roots remain the same, but we have an added *getsegment*() that rotates the segment through the appropriate angle. By setting *oldgain*[1] to zero before the start of each row, we ensure that no false segment is plotted. The effect is shown in Figure 12.4.

It is a little ragged around the "breakaway point" on the real axis, but will certainly serve its purpose.

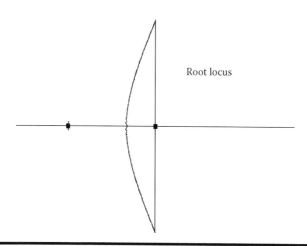

Figure 12.4   Root locus plot of $1/s(s+1)^2$**. (Screen grab from www.esscont. com/12/rootlocus.htm)**

12.4 Plotting with Poles and Zeroes

With the easy availability of computer power, the ability to plot a root locus by hand might seem have dubious value. However, a root locus is an easy subject for an examination question!

It is also useful to have some insight to allow us to verify any computer solution.

Although we have talked of poles and zeros, the only systems we have looked at so far have only had poles. Let us consider a system with one of each,

$$G(s) = \frac{s+1}{s+2}$$

Now the closed loop gain is

$$\frac{kG}{1+kG}$$

$$= \frac{k\dfrac{s+1}{s+2}}{1+k\dfrac{s+1}{s+2}}$$

$$= \frac{k(s+1)}{(1+k)s+(2+k)}$$

Clearly the closed loop system has a zero in exactly the same place as the open loop. This is not surprising. If G gives zero output for some particular value of complex input frequency, then no amount of feedback will make it change its mind.

There is just one pole, at value $-(2+k)/(1+k)$. Now if k is small, the pole is near its open loop position. As k increases, the pole moves along the axis until at very large k it approaches the position of the zero at -1.

Now let us try two poles and one zero,

$$\frac{s+1}{s(s+2)}$$

This time the closed loop gain is

$$\frac{k\dfrac{s+1}{s(s+2)}}{1+k\dfrac{s+1}{s(s+2)}}$$

$$= \frac{k(s+1)}{s^2+s(2+k)+k}$$

so the poles are at the roots of

$$s^2 + (2+k)s + k = 0$$

i.e.,

$$s = -1 - k/2 \pm \sqrt{(1+k/2)^2 - k}$$

$$= -1 - k/2 \pm \sqrt{1 + k^2/4}$$

Both roots are real. For small k the roots are near -2 and 0, where we would expect. As k tends to infinity, the value of the square root tends toward $k/2$, and we find one root near -1 while the other is dashing off along the negative real axis.

What have we learned from these three simple examples?

1. The closed loop system has the same zeros as the open loop.
2. As k tends to infinity, a pole approaches each zero.
3. If we have two poles and one zero, the "spare" pole heads off along the negative real axis.
4. If we have two poles and no zeros, so that both are "spare," then they head off North and South in the imaginary directions.

Can these properties be generalized to a greater number of poles and zeros, and what other techniques can be discovered?

12.5 Poles and Polynomials

We can represent the gain of the sort of linear system that we have been considering as a ratio of two polynomials in the form

$$G(s) = a\frac{Z(s)}{P(s)}$$

Here $P(s)$ and $Z(s)$ are defined by n poles and m zeroes as

$$P(s) = (s - p_1)(s - p_2)\dots(s - p_n)$$

and

$$Z(s) = (s - z_1)(s - z_2)\dots(s - z_m)$$

allowing the closed loop gain to be written as

$$\frac{kG}{1+kG} = \frac{kaZ(s)}{P(s) + kaZ(s)}$$

To plot the progress of the poles, we must look at the roots of

$$P(s) + kaZ(s) = 0$$

Without losing generality we can absorb the constant a into the value of k. When we multiply out $P(s)$ and $Z(s)$, the coefficient of the highest power of s in each of them is unity.

We can draw a number of different conclusions according to whether $m = n$, $m = n - 1$, or $m < n - 1$. We will rule out the possibility of having more zeros than poles, since that would imply that the gain increased indefinitely with increasing frequency.

In every case, when $k = 0$ the poles start from their open loop locations.

If there are equal numbers of poles and zeros, the final resting place of each pole is on a zero, since we can divide through by k and see the $P(s)/k$ contribution to the coefficients disappear as k becomes infinite.

If $m = n - 1$, then the leading term is unchanged by k. Again we can divide through by k, to get an expression that for very large k approximates to $s^n/k + Z(s) = 0$. The poles head for the zeros, while the "spare" pole heads for minus infinity along the real axis.

If $m < n - 1$, the case is more interesting. We have seen a pair of poles heading out along north and south asymptotes. Some algebraic gymnastics can prove that if we have three "spare" poles they will move out in the directions of the cube root of -1. They can be seen setting out in Figure 12.4. Four extra poles will move out in four diagonal directions, and so on.

What is of interest is the point where these asymptotes intersect.

We know that m of the roots will head for the roots of $Z(s)$, but what of the rest? We can use a sort of long division of the polynomials to express

$$P(s) = Q(s)Z(s) + R(s)$$

where the remainder $R(s)$ is of a lower order than $Z(s)$. So our denominator is

$$P(s) + kZ(s)$$

$$= Q(s)Z(s) + R(s) + kZ(s)$$

$$= (Q(s) + k)Z(s) + R(s)$$

As k gets very large, the coefficients of R will be swamped by those of kZ and we will be left with poles near the zeroes and near the roots of $Q(s) + k = 0$.

What do we know about the coefficients of Q?

In the long division, the first term will be s^{n-m}. The coefficient of the second term in s^{n-m-1} will be obtained by subtracting the coefficient of s^{m-1} in Z from the coefficient of s^{n-1} in P. But these coefficients represent the sum of the roots of Z and the sum of the roots of *P*, while this new coefficient will represent the sum of the spare poles.

So we can say that as *k* becomes large, the sum of the roots of the "spare poles" will be the difference between the sum of the open loop poles and the sum of the open loop zeros.

Another way to say this is:

Give each open loop pole a "weight" of $+1$. Give each zero a weight of -1. Then the asymptotes will meet at the "center of gravity."

Q 12.5.1

Show that the root locus of the system $1/s(s+1)^2$ has three asymptotes which intersect at $s=-2/3$. Make a very rough sketch.

Q 12.5.2

Add a zero at $s=-2$, so that the system becomes $(s+2)/s(s+1)^2$. What and where are the asymptotes now?

Q 12.5.3

A "zoomed out" version of the root-locus plotter is to be found at www.esscont. com/12/rootzoom.htm.

Edit the values of the poles and the zeros to test the assertions of this last section. Figure 12.5 shows the plot for a zero at -2 and poles at 0 and -1.

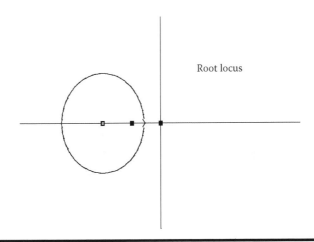

Figure 12.5 Screen grab of www.esscont.com/12/rootzoom.htm, for $G=(s+2)/s(s+1)$.

There are more rules that can be derived for plotting the locus by hand.

It can be shown that at a "breakaway point" where the poles join and split away in different directions, then the derivative $G'(s) = 0$.

It can be shown that those parts of the real axis that have an odd number of poles or zeroes on the axis to the right of them will form part of the plot.

But it is probably easier to make use of the root-locus plotting software on the website.

One warning is that some operating systems will put up an error message if the JavaScript is kept busy for more than five seconds. The plot can be made much neater by reducing ds to 0.2, but reducing it to 0.01 might provoke the message.

12.6 Compensators and Other Examples

We have so far described the root locus as though it were only applicable to unity feedback. Suppose that we use some controller dynamics, either at the input to the system or in the feedback loop (see Figure 12.6).

Although the closed loop gains are different, the denominators are the same. The root locus will be the same in both cases, with the poles and zeroes of system and controller lumped together.

The root locus can help not merely with deciding on a loop gain, but in deciding where to put the roots of the controller.

Q 12.6.1

An undamped motor has response $1/s^2$. With a gain k in front of the motor and unity feedback around the loop, sketch the root locus. Does it look encouraging?

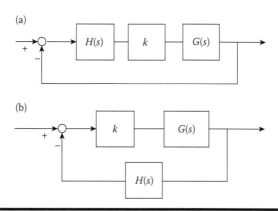

Figure 12.6 Two configurations with dynamics in the feedback loop (a) dynamics at the system input (b) dynamics in the feedback path.

Q 12.6.2

Now apply phase advance, by inserting $H(s)=(s+1)/(s+3)$ in front of the motor. Does the root locus look any more hopeful?

Q 12.6.3

Change the phase advance to $H(s)=(3s+1)/(s+3)$.

Let us work out these examples here. The system $1/s^2$ has two poles at the origin. There are two excess poles so there are two asymptotes in the positive and negative imaginary directions. The asymptotes pass through the "center of gravity," i.e., through $s=0$. No part of the real axis can form part of the plot, since both poles are encountered together.

We deduce that the poles split immediately, and make off up and down the imaginary axis. For any value of negative feedback, the result will be a pair of pure imaginary poles representing simple harmonic motion.

Now let us add phase advance in the feedback loop, with an extra pole at $s=-3$ and a zero at $s=-1$. There are still two excess poles, so the asymptotes are still parallel to the imaginary axis. However they will no longer pass through the origin.

To find their intersection, take moments of the poles and zero. We have contribution 0 from the poles at the origin, −3 from the other pole and +1 from the zero. The total, −2, must be divided by the number of excess poles to find the intersection, at $s=-1$.

How much of the axis forms part of the plot? Between the pole at −3 and the zero, there is one real zero plus two poles to the right of s—an odd total. To the left of the single pole and to the right of the zero the total is even, so these are the limits of the part of the axis that forms part of the plot.

Putting all these deductions together, we could arrive at a sketch as shown in Figure 12.7. The system is safe from instability. For large values of feedback gain, the resonance poles resemble those of a system with added velocity feedback.

Now let us look at example Q 12.6.3. It looks very similar in format, except that the phase advance is much more pronounced. The high-frequency gain of the phase advance term is in fact nine times its low frequency value.

We have two poles at $s=0$ and one at $s=-3$, as before. The zero is now at $s=-1/3$. For the position of the asymptotes, we have a moment −3 from the lone pole and +1/3 from the zero. The asymptotes thus cut the real axis at half this total, at −4/3.

As before, the only part of the real axis to form part of the plot is that joining the singleton pole to the zero. It looks as though the plot may be very similar to the last.

Some calculus and algebra, differentiating $G(s)$ twice, would tell us that there are breakaway points on the axis, a three-way split. With the loop gain $k=3$ we have three equal roots at $s=-1$ and the response is very well damped indeed. By all means try this as an exercise, but it is easier to look at Figure 12.8.

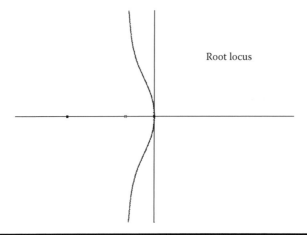

Figure 12.7   Two poles at the origin, compensator has a pole at −3 and a zero at −1. (Screen grab from www.esscont.com/12/rootzoom2.htm)

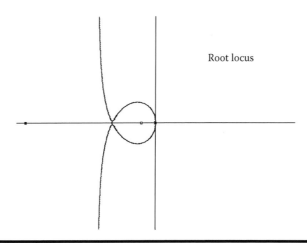

Figure 12.8   Two-integrator system, with compensator pole at −3 and zero at −1/3.

12.7 Conclusions

The root locus gives a remarkable insight into the selection of the value of a feedback parameter. It enables phase advance and other compensators to be considered in an educated way. It can be plotted automatically by computer, or with only a little effort by hand by the application of relatively simple rules.

Considerable effort has been devoted here to this technique, since it is effective for the analysis of sampled systems too. It has its restrictions, however.

The root locus in its natural form only considers the variation of a single parameter. When we have multiple inputs and outputs, although we can still consider a single characteristic equation we have a great variety of possible feedback arrangements. The same set of closed loop poles can sometimes be achieved with an infinite variety of feedback parameters, and some other basis must be used for making a choice. With multiple feedback paths, the zeroes no longer remain fixed, so that individual output responses can be tailored. Other considerations can be non-linear ones of drive saturation or energy limitation.

Chapter 13

Fashionable Topics in Control

13.1 Introduction

It is the perennial task of researchers to find something new. As long as one's academic success is measured by the number of publications, there will be great pressure for novelty and abstruseness. Instead, industry's real need is for the simplest controller that will meet all the practical requirements.

Through the half century that I have been concerned with control systems, I have seen many fashions come and go, though some have had enough substance to endure. No doubt many of the remarks in this chapter will offend some academics, but I hope that they will still recommend this book to their students. I hope that many others will share my irritation at such habits as giving new names and notation to concepts that are decades old.

Before chasing after techniques simply because they are novel, we should remind ourselves of the purpose of a control system. We have the possibility of gathering all the sensor data of the system's outputs. We can also accumulate all the data on inputs that we have applied to it. From this data we must decide what inputs should be applied at this moment to cause the system to behave in some manner that has been specified.

Anything else is embroidery.

13.2 Adaptive Control

This is one of the concepts with substance. Unfortunately, like the term "Artificial Intelligence," it can be construed to mean almost anything you like.

In the early days of autopilots, the term was used to describe the modification of controller gain as a function of altitude. Since the effectiveness of aileron or elevator action would be reduced in the lower pressure of higher altitudes, "gain scheduling" could be used to compensate for the variation.

But the dream of the control engineer was a black box that could be wired to the sensors and actuators and which would automatically learn how best to control the system.

One of the simpler versions of this dream was the *self-tuning regulator.*

Since an engineer is quite capable of adjusting gains to tailor the system's performance, an automatic system should be capable of doing just as well. The performance of auto-focus systems in digital video cameras is impressive. We quite forgive the flicker of blurring that occasionally occurs as the controller *hill-climbs* to find the ideal setting. But would a twitching autopilot be forgiven as easily?

In philosophical terms, the system still performs the fundamental task of a control system as defined in the introduction. However, any expression for the calculation of the system input will contain products or other nonlinear functions of historical data, modifying the way that the present sensor signals are applied to the present inputs.

13.3 Optimal Control

Some magical properties of optimal controllers define them to be the "best." This too has endured and the subject is dealt with at some length in Chapter 22. However, the quality of the control depends greatly on the criterion by which the response is measured. A raft of theory rests on the design of linear control systems that will minimize a *quadratic cost function*. All too often, the cost function itself is designed with no better criterion than to put the poles in acceptable locations, when *pole assignment* would have performed the task in a better and more direct way.

Nevertheless there is a class of *end point* problems where the control does not go on forever. Elevators approach floors and stop, aeroplanes land automatically, and modules land softly on the Moon. There are pitfalls when seeking an absolute minimum, say of the time taken to reach the next traffic light or the fuel used for a lunar descent, but there are suboptimal strategies to be devised in which the end point is reached in a way that is "good enough."

13.4 Bang–Bang, Variable Structure, and Fuzzy Control

Recognizing that the inputs are constrained, a *bang–bang* controller causes the inputs to take extreme values. As described in Section 6.5, rapid switching in a *sliding mode* is a feature of *variable structure* control. The sliding action can reduce the effective order of the system being controlled and remove the dependence of the performance on some of the system parameters. Consider for example a *bang–bang velodyne loop* for controlling the speed of a servomotor.

A tachometer measures the speed of the motor and applies maximum drive to bring the speed to the demanded value. When operating in sliding mode, the drive switches rapidly to keep the speed at the demanded value. To all intents and purposes the system now behaves like a first-order one, as long as the demand signal does not take the operation out of the sliding region. In addition, the dynamics will not depend on the motor gain in terms of acceleration per volt, although this will obviously determine the extent of the sliding region.

Variable structure control seems to align closely with our pragmatic approach for obtaining maximum closed loop stiffness. However, it seems to suffer from an obsessive compulsion to drive the control into sliding.

When we stand back and look at the state-space of a single-constrained-input system, we can see it break into four regions. In one region we can be certain that the drive must be a positive maximum, such as when position and velocity are both negative. There is a matching region where the drive must be negative. Close to a stationary target we might wish the drive to be zero, instead of switching to and fro between extremes. That leaves a fourth region in which we have to use our ingenuity to control the switching.

Simulation examples have shown us that when the inputs are constrained, a nonlinear algorithm can perform much better than a linear one. "Go by the book" designers are therefore attracted by any methodology that formalizes the inclusion of nonlinearities. In the 1960s, advanced analog computers possessed a "diode function generator." A set of knobs allowed the user to set up a piecewise-linear function by setting points between which the output was interpolated.

Now the same interpolated function has re-emerged as the heart of *fuzzy control*. It comes with some pretentious terminology. The input is related to the points where the gradient changes by a *fuzzifier* that allocates *membership* to be shared between *sets* of neighboring points. Images like Figure 13.1 appear in a multitude of papers. Then the output is calculated by a *defuzzifier* that performs the interpolation. This method of constructing a nonlinear output has little wrong with it except the jargon. Scores of papers have been based on showing some improved performance over linear control.

Another form of fuzzy *rule based* control results from inferior data. When reversing into a parking space, relying on helpful advice rather than a rear-view camera, your input is likely to be "Plenty of room" followed by "Getting close" and finally "Nearly touching." This is fuzzy data, and you can do no better than base your control on simple rules. If there is a sensor that gives clearance accurate to a millimeter, however, there is little sense in throwing away its quality to reduce it to a set of fuzzy values.

Bang-bang control can be considered as an extreme form of a fuzzy output, but by modulating it with a *mark-space* ratio the control effect can be made linear.

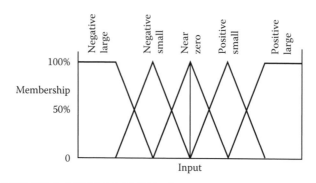

Figure 13.1   A fuzzifier.

13.5 Neural Nets

When first introduced, the merit of *neural nets* was proclaimed to be their massive parallelism. Controllers could be constructed by interconnecting large numbers of simple circuits. These each have a number of inputs with variable *weighting functions*. Their output can switch from one extreme to another according to the weighted sum of the inputs, or the output can be "softened" as a *sigmoid* function.

Once again these nets have the advantage of an ability to construct nonlinear control functions. But rather than parallel computation by hardware, they are likely to be implemented one-at-a-time in a software simulation and the advantage of parallelism is lost.

There is another side to neural nets, however. They afford a possibility of adaptive control by manipulating the weighting parameters. The popular technique for adjusting the parameters in the light of trial inputs is termed *back propagation*.

13.6 Heuristic and Genetic Algorithms

In 1935, Ross Ashby wrote a book called "Design for a brain." Of course the title was a gross overstatement. The essence of the book concerned a feedback controller that could modify its behavior in the light of the output behavior to obtain *hyperstability*. If oscillation occurred, the "strategy" (a matter of simple circuitry) would switch from one preset feedback arrangement to the next. Old ideas do not die.

Heuristic control says, in effect, "I do not know how to control this," then tries a variety of strategies until one is found that will work. In a *genetic algorithm*, the fumbling is camouflaged by a smokescreen of biologically inspired jargon and pictures of double-helix chromosomes. The paradigm is that if two strategies can be found that are successful, they can be combined into a set of "offspring" of which one might perform better. Some vector encryption of the control parameters is termed a *chromosome* and random combinations are tested to select the best.

It is hard to see how this can be better than deterministic *hill climbing*, varying the control parameters systematically to gain a progressive improvement.

All too often these methods suffer from the deficiency that adaptation is determined by a set of simulated tests, rather than real experimental data. Something that works perfectly in simulation can fall apart when applied in practice.

13.7 Robust Control and *H*-infinity

From the sound of it, *robust control* suggests a controller that will fight to the death to eliminate disturbances. The truth is very different. The "robustness" is the ability of the system to remain stable when the gain parameters vary. As a result, control is likely to be "soft."

One of the fashionable design techniques for a robust system has been "*H-infinity*." A system becomes unstable when the loop gain is unity. If we can choose the feedback so that there is a limit on the magnitude of the gain, assessed over all frequencies, then instability can be avoided. Remember that the systems in question are multi-input and multi-output (MIMO) so the feedback choice is not trivial.

Several decades ago, in the quest to simplify feedback for MIMO systems, one suggestion was *dyadic feedback*. The output signals could be mixed together into a single feedback path, then this signal could be shared out among the various inputs. As a result, although the rank of the feedback matrix is just unity, it is possible to assign the values of the closed loop poles. Unfortunately the closed loop zeros can be less than desirable.

13.8 The Describing Function

This is another technique that has endured. A long-known problem has been the determination of stability when there is a nonlinearity in the system. When a system oscillates, the loop gain is exactly unity, as shown in Figure 13.2.

Let us state the obvious: When we close the loop, the feedback signal arriving at the input is exactly the same as the input that produces it. To the signal at the input, the loop gain is exactly one, by any reckoning, whether it is a decaying exponential or a saturated square-wave oscillation. Once we allow the system to become nonlinear, the "eigenfunction" is no longer a simple (or complex!) exponential, but can take a variety of distorted forms.

To put a handle onto the analysis of such a function, we must make some assumptions and apply some limitations. We can look at such effects as clipping, friction and backlash, and we can assume that the oscillation that we are guarding against is at least approximately sinusoidal.

With the assumption that the oscillation signal is one that repeats regularly, we open up the possibility of breaking the signal into a series of sinusoidal components

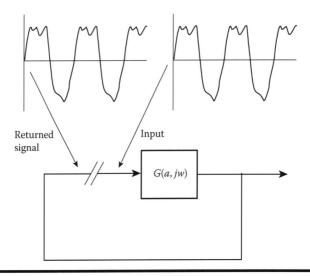

Figure 13.2 Signals in an oscillator.

by calculating its Fourier series. The fundamental sinewave component of the signal entering the system must be exactly equal to the fundamental component of the feedback, and so we can at least start to build up an equation.

We can consider the application of sinewaves of varying amplitudes to the system, as well as of varying frequencies, and will extract a fundamental component from the feedback which is now multiplied by a gain function of both frequency and amplitude, the describing function of the system $G(a, j\omega)$. As ever, we are concerned with finding if G can take the value −1 for any combination of frequency and amplitude.

Of course, the method depends on an approximation. It ignores the effect of higher harmonics combining to recreate a signal at the fundamental frequency. However, this effect is likely to be small. We can use the method both to estimate the amplitude of oscillation in an unstable system that has constraints, and to find situations where an otherwise stable system can be provoked into a limit cycle oscillation.

13.9 Lyapunov Methods

In an electronic controller, a sharp nonlinearity can occur as an amplifier saturates. In the world at large, we are lucky to find any system that is truly linear. The expression of a system in a linear form is nearly always only an approximation to the truth. Local linearization is all very well if we expect the disturbances to be small, but that will often not be the case. The phase-plane has been seen to be useful in examining piecewise-linear systems, and in some cases it is no doubt possible to find isoclines for more general nonlinearities. However, we would like to find a method for analyzing the stability of nonlinear systems in general, including systems of higher order than two.

One long established approach is the "direct" method of *Lyapunov*, astonishingly simple in principle but sometimes needing ingenuity to apply. First, how should we define stability?

If we disturb the system, its state will follow a trajectory in n-dimensional state space. If all such trajectories lead back to a single point at which the system comes to rest, then the system is *asymptotically* stable. If some trajectories diverge to infinity, then the system is unstable.

There is a third possibility. If all trajectories lead to a bounded region of the state space, remaining thereafter within that region without necessarily settling, then the system is said to have *bounded stability*.

These definitions suggest that we should examine the trajectories, to see whether they lead "inward" or "outward"—whatever that might mean. Suppose that we define a function of the state, $L(\mathbf{x})$, so that the equation $L(\mathbf{x}) = r$ defines a closed "shell." (Think of the example of circles or spheres of radius r.) Suppose that the shell for each value of r is totally enclosed in the shell for any larger value of r. Suppose also that as r is reduced to zero so the shells converge to a single point of the state space.

If we can show that on any trajectory the value of r continuously decreases until r becomes zero, then clearly all trajectories must converge. The system is asymptotically stable.

Alternatively, if we can find such a function for which r increases indefinitely, then the system is unstable.

If r aims for some range of non-zero values, reducing if it is large but increasing if small, then there is a limit cycle defined as bounded stability. The skill lies in spotting the function L.

Q 13.9.1

Find a suitable Lyapunov function to analyze the nonlinear system

$$\ddot{x} + \dot{x}(x^2 + \dot{x}^2 - 1) + x = 0$$

13.10 Conclusion

These and many other techniques will continue to fill the journal pages. They will also fill the sales brochures of the vendors of "toolboxes" for expensive software packages. Some methods might be valuable innovations. Many more, however, will recycle old and tried concepts with novel jargon. The authors of all too many papers are adept at mathematical notation but have never applied control to any practical system other than a simplified laboratory experiment. Even for this, they have probably used a "toolbox" that just involves dragging and dropping icons.

The engineer might explain that a system has four state variables. When you come upon a paper with an insistence that the state $\mathbf{x} \in \Re^4$, you will recognize its author as someone who is trying to impress rather than enlighten.

Chapter 14

Linking the Time and Frequency Domains

14.1 Introduction

The state-space description of a system can come close to describing its ingredients; state equations can be built from the physical relationships between velocities and positions, levels, flows, and any other identifiable quantities. The frequency domain description is more concerned with responses that can be measured at the outputs when signals applied to the inputs, regardless of what may be contained in the "black box." We have looked at aspects of both techniques in isolation; now let us tie the two areas together.

14.2 State-Space and Transfer Functions

When considering a position control system, we dithered between state equations and second order differential equations and switched fairly easily between the two. A second order differential equation, such as

$$\ddot{y} + 5\dot{y} + 6y = 6u$$

expresses a clear relationship between the single input u, the position demand, and the output position y of the motor. It is the work of a moment to sprinkle a flavoring of Laplace onto this equation and start to apply any of the methods of the root locus.

A set of state equations describe exactly the same system, but does so in formalized terms that tie position and velocity into simultaneous first order equations, and express the output as a mixture of these state variables:

$$\begin{bmatrix} \dot{x}_1 \\ \dot{x}_2 \end{bmatrix} = \begin{bmatrix} 0 & 1 \\ -6 & -5 \end{bmatrix} \begin{bmatrix} x_1 \\ x_2 \end{bmatrix} + \begin{bmatrix} 0 \\ 6 \end{bmatrix} u$$

$$y = \begin{bmatrix} 1 & 0 \end{bmatrix} \begin{bmatrix} x_1 \\ x_2 \end{bmatrix} \qquad (14.1)$$

This last equation does no more than select the motor position as our output, yet involves multiplying a matrix and a vector. This formal method is most appealing when we have a computer at our fingertips; a computer will perform a vast number of matrix multiplications in less time than it takes for us to be ingenious.

With a single input and a single output, we find a single transfer function linking the two, just as we would expect. If we have two inputs and one output, and if the system is linear, then the output will be made up of the sum of the effects of the two inputs. We can find two transfer functions, one associated with each input, allowing the output to be expressed as

$$Y(s) = G_1(s).U_1(s) + G_2(s).U_2(s).$$

In the same way, one input and two outputs will also have two transfer functions.

When we move up to two inputs and two outputs, we discover four transfer functions. Each output is linked to each of the two inputs, and this can be represented most neatly in matrix form

$$\begin{bmatrix} Y_1(s) \\ Y_2(s) \end{bmatrix} = \begin{bmatrix} G_{11}(s) & G_{12}(s) \\ G_{21}(s) & G_{22}(s) \end{bmatrix} \begin{bmatrix} U_1(s) \\ U_2(s) \end{bmatrix}$$

where each of the G's will probably be a ratio of two polynomials in s. To gain the greatest advantage from the computer, we would like to find routine ways of switching between state-space and transfer function forms.

14.3 Deriving the Transfer Function Matrix

We wish to find a transfer function representation for a system given in state-space form. We start with

$$\dot{\mathbf{x}} = \mathbf{A}\mathbf{x} + \mathbf{B}\mathbf{u}$$
$$\mathbf{y} = \mathbf{C}\mathbf{x}$$

(14.2)

Since we are looking for a transfer function, it seems a good idea to take the Laplace transform of $\mathbf{x}$ and deal with $\mathbf{X}(s)$. In Section 10.5 we saw that

$$\mathcal{L}\{f'(t)\} = sF(s)f(0)$$

which means that

$$\mathcal{L}\{\dot{\mathbf{x}}\} = s\mathbf{X}(s) - \mathbf{x}(0)$$

(14.3)

so

$$s\mathbf{X}(s) - \mathbf{x}(0) = \mathbf{A}\mathbf{X}(s) + \mathbf{B}\mathbf{u}(s)$$

i.e.,

$$s\mathbf{X}(s) - \mathbf{A}\mathbf{X}(s) = \mathbf{B}\mathbf{U}(s) + \mathbf{x}(0)$$

Now we can mix the two terms on the left together more easily if we write

$$s\mathbf{X}(s)$$

as

$$s\mathbf{I}\mathbf{X}(s)$$

where $\mathbf{I}$ is the unit matrix.

This enables us to take out $\mathbf{X}(s)$ as a factor, to obtain

$$(s\mathbf{I} - \mathbf{A})\mathbf{X}(s) = \mathbf{B}\mathbf{U}(s) + \mathbf{x}(0)$$

(14.4)

All that remains is to find the inverse of the matrix $(s\mathbf{I}-\mathbf{A})$, and we can multiply both sides by it to obtain a clear $\mathbf{X}(s)$ on the left. Remember that the order of multiplication is important. The inverse must be applied in front of each side of the equation to give

$$\mathbf{X}(s) = (s\mathbf{I} - \mathbf{A})^{-1}(\mathbf{B}\mathbf{U}(s) + \mathbf{x}(0))$$

We also have

$$Y(s) = CX(s)$$

so the outputs are related to the inputs by

$$Y(s) = C(sI - A)^{-1}(BU(s) + x(0))$$

For the transfer functions, we are not really interested in the initial conditions of x; we assume that they are zero when the input is applied, or that a sinusoidal output has settled to a steady state. The equation can be reduced to

$$Y(s) = C(sI - A)^{-1}BU(s). \tag{14.5}$$

Now the matrix

$$C(sI - A)^{-1}B$$

is a matrix of transfer functions having as many rows as there are outputs, and as many columns as there are inputs.

To invert a matrix, we first form the matrix of the "cofactors." The elements of this matrix are the determinants of what is left of the original matrix when we delete the row and column corresponding to the position in the new matrix. We multiply these by plus or minus 1, in a $+ - + -$ pattern, then divide the whole thing by the determinant of the original matrix.

When we evaluate $\det(sI{-}A)$, we will get a polynomial in s. This is called the characteristic polynomial of A, and it then appears in the denominator of every term of the inverse.

In other words, the roots of the characteristic polynomial provide the poles of the system as it stands, before the application of any extra feedback. Do not expect that each of the transfer function terms will have all these poles; in many cases they will cancel out with factors of the numerator polynomials. However, we can assert that the poles of the transfer functions can only come from the roots of the characteristic equation.

Q 14.3.1

Use this method to derive a transfer function for the example defined by Equations 14.1.

Q 14.3.2

Derive the transfer function matrix for

$$\begin{bmatrix} \dot{x}_1 \\ \dot{x}_2 \end{bmatrix} = \begin{bmatrix} -3 & 0 \\ 0 & -2 \end{bmatrix} \begin{bmatrix} x_1 \\ x_2 \end{bmatrix} + \begin{bmatrix} 1 & 0 \\ 0 & 1 \end{bmatrix} \begin{bmatrix} u_1 \\ u_2 \end{bmatrix}$$

$$y = \begin{bmatrix} 1 & 1 \end{bmatrix} \begin{bmatrix} x_1 \\ x_2 \end{bmatrix}$$

(14.6)

Q 14.3.4

Derive the transfer matrix for

$$\begin{bmatrix} \dot{x}_1 \\ \dot{x}_2 \end{bmatrix} = \begin{bmatrix} -3 & 0 \\ 0 & -2 \end{bmatrix} \begin{bmatrix} x_1 \\ x_2 \end{bmatrix} + \begin{bmatrix} -1 \\ 1 \end{bmatrix} u$$

$$y = \begin{bmatrix} 6 & 6 \end{bmatrix} \begin{bmatrix} x_1 \\ x_2 \end{bmatrix}$$

(14.7)

14.4 Transfer Functions and Time Responses

When we looked at the Laplace transform, we saw a set of functions of time that could be transformed into functions of *s*. The output of a system in the form of a function of s was simply given by multiplying the transform of the input function by the transfer function. But an important question to ask is, "What function of time has a Laplace transform of just 1?"

The Laplace transform of the "unit step," which is zero for all negative time and 1 for all positive time will be

$$\int_0^\infty 1.e^{-st}\,dt$$

$$= \frac{1}{s}$$

So since an integration corresponds to an extra $1/s$ in the denominator, we are looking for a function that when integrated will give the unit step. It is the "unit impulse," infinitesimally thin and infinitely high! (Figure 14.1)

The unit impulse has an area of 1. It is the limit of shrinking the width to zero while keeping the area at unity.

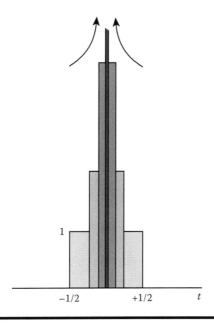

Figure 14.1   The unit impulse.

So the transfer function of a system corresponds to the Laplace transform of its output, when the input is the unit impulse. For any given transfer function $H(s)$ we can find a corresponding function of time $h(t)$ which is the "impulse response" of the system.

Suppose that our system is defined by $H(s) = 1/(s + a)$. A trip to the tables of Laplace transforms will tell us that $h(t) = e^{-at}$, in other words the initial response at $t = 0$ has a value of unity, decaying to zero with time constant $1/a$. Is there some way of gaining insight into the response function without actually looking up or working out the inverse transform?

There are two powerful properties of the Laplace transform which will help us, the "initial value theorem" and the "final value theorem."

The initial value theorem tells us that

$$\lim_{t \to 0} f(t) = \lim_{s \to \infty} sF(s)$$

We have to construct this limit for positive values of t, decreasing toward zero, since there might be a step change in $f(t)$ at $t = 0$.

The final value theorem states that

$$\lim_{t \to \infty} f(t) = \lim_{s \to 0} sF(s)$$

Rather than construct a formal proof, let us look for a plausible argument. If $F(s)$ is a ratio of polynomials, we can rearrange it as a ratio of polynomials in $1/s$. We can then perform a "long division" to obtain a power series in $1/s$. Thus, $1/(s+1)$ would become $(1/s)/(1+1/s)$, which we would divide out to obtain

$$\frac{1}{s} - \frac{1}{s^2} + \frac{1}{s^3}\cdots$$

The inverse transform of this series will be a power series in t, since $1/s^{n+1}$ is the Laplace transform of $t^n/n!$ If we are only interested in the value at $t = 0$, we can ignore all terms beyond the $1/s$ term. Now if there is a constant at the start of the series, we have an infinite impulse on our hands. If it starts with a $1/s$ term, we have a respectable step, while if the first term is of higher power then our initial value is zero.

To extract the $1/s$ term's value, all we need to do is multiply the series by s, and let s tend to infinity. If there is a constant term, we will correctly get an infinite result. But why go to the trouble of dividing out? We need only multiply $F(s)$ by s, let s tend to infinity, and there is the initial value of $f(t)$.

The next argument, to demonstrate the final value theorem, is even more woolly!

If we have a relationship $Y(s) = H(s)U(s)$, and if $H(s)$ is the ratio of two polynomials in s, $Z(s)/P(s)$, then we have

$$P(s)Y(s) = Z(s)U(s)$$

This relationship can be construed as meaning that $y(t)$ and $u(t)$ are linked by a differential equation

$$P\left(\frac{d}{dt}\right)y(t) = Z\left(\frac{d}{dt}\right)u(t)$$

If we assume that everything is stable, and that the $u(t)$ which is applied is the unit step, then eventually all the derivatives of y will become zero, while the derivatives of u will have long passed after provoking the initial transients. Thus for the limiting value of y as t tends to infinity we can let d/dt tend to zero in the polynomials. This is just the same as letting s tend to zero, so we can reassemble $H(s)$ to state that:

If a unit step is applied to a transfer function $H(s)$, then as time tends to infinity the output will tend to

$$\lim_{s \to 0} H(s)$$

Now the unit impulse is the time-derivative of the unit step, so (begging all sorts of questions of convergence) the output for an impulse input will be the derivative of the output for a step input. We can serve up a derivative by multiplying $H(s)$ by another s, obtaining

$$\lim_{t \to \infty} f(t) = \lim_{s \to 0} sF(s)$$

Well I warned you that the argument was woolly!

In fact it is often more helpful to look at the step response of a filter than at the impulse response. Consider the phase-advance element with transfer function

$$H(s) = \frac{1+3s}{1+s}$$

To find the initial and final values resulting from applying a step to the input, we must look at the result of applying $1/s$ to the transfer function $H(s)$. Instead of taking the limits of $sH(s)$, the input $1/s$ will cancel the "spare" s to leave us with the limits of $H(s)$ as s tends to infinity and as s tends to zero, respectively.

For the initial value of the step response, we let s tend to infinity in $H(s)$, which in this case gives us the value 3. For the final value of the step response, we let s tend to zero in $H(s)$, here giving value 1. From the denominator we expect the function e^{-t} to be involved, so we can intelligently sketch the step response as in Figure 14.2.

We can add step responses together to visualize the results of various transfer functions. The phase advance we have just considered, for instance, can be split into two terms

$$\frac{1+3s}{1+s} = 1 + \frac{2s}{1+s}$$

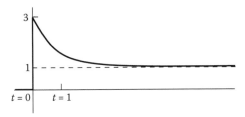

Figure 14.2 Step response of (1 + 3s)/(1 + s).

We can generate a filter to produce this response by mixing a "straight-through" term having a gain of unity with another signal that gives the second term above. Now this second term's step response has initial value 2 and final value 0. To construct the phase advance we need to add this transient to the basic step.

The "phase advance" circuit of Figure 14.3 will give a transfer function $-(1 + 3s)/(1 + s)$.

In passing, note that in the simple circuit of Figure 14.4 the resistor voltage is filtered by $s/(1 + s)$ while the capacitor voltage is filtered by $1/(1 + s)$, a "lag" which smooths the response by attenuating higher frequencies. When these responses are added together, we get back to the original input. In other words, we can construct the transient response by subtracting a lagged signal from the original signal.

14.5 Filters in Software

Earlier on, we saw that a dynamic system could be simulated by means of its state equations. When a signal is applied to the system, an output is obtained which is related to the input by means of a transfer function. When we apply the same input to our simulation, if all is well we will obtain the same output function as in the

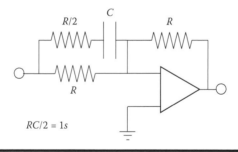

Figure 14.3 Phase advance circuit.

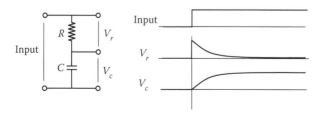

Figure 14.4 Voltage waveforms making up the step response of an RC circuit.

real system. We touched on this in a practical sense in Chapter 9, but now we can treat the subject more rigorously.

A filter is merely the means of applying a transfer function to an input signal. It might be constructed from resistors, capacitors, and amplifiers or it might contain pneumatic reservoirs. The implementation is of secondary concern; it is the effect on the input signal that matters.

If we simulate such a filter, we have not merely made a simulation, we have constructed an alternative filter which we could use in place of the original (assuming of course that the input and output signals have the right qualities).

Let us look yet again at the task of constructing the phase-advance element $(1 + 3s)/(1 + s)$, this time using a few lines of software in a computer program.

The first task is to set up some state equations. The denominator suggests that all we need is

$$\dot{x} = -x + u$$

to give us the "lag" $X(s) = U(s)/(s + 1)$. When we try to set up an output in the form

$$y = \mathbf{C}\,\mathbf{x}$$

However, we immediately find problems. The state vector only has one element, which gives a lagged version of the input. No amount of multiplication by constants will sharpen it into a phase-advance. Remember however that a transient can be obtained by subtracting a lag from the original input. If we look at

$$y = 3u - 2x$$

we see that $Y(s) = (3 - 2/(1 + s))U(s)$, which reduces to the desired function.

To achieve a transfer function with as many zeros as it has poles, it is clear that we must broaden our general definition of the output of a system to

$$y = \mathbf{C}\,x + \mathbf{D}\,u$$

Why was the output not defined in this way in the first place? The reason is that it allows the output to change instantly when the input changes. So when we come to apply feedback around a system that has an "instant response," we will have to solve simultaneous algebraic equations as well as differential equations to find the output. To produce generalized theorems, it is safer to stick to the simpler form of output, and to treat this form as an exception whenever it cannot be avoided.

So now we have two equations, where the rate-of-change of *x* can be represented by a variable named *dx*. These can be calculated by the computer assignment statements:

```
dx = u - x
y = 3*u-2*x
```

We can introduce the time element by the simplest of integrations

```
x = x + dx*dt
```

and all the ingredients are present for the software model. We might choose a time-step of 0.01 seconds to arrive at:

```
dx = u  -  x
x = x + dx*.01
y = 3*u  -  2*x
```

For *dt* = 0.01, if we can be sure that these lines will be executed 100 times per second, then we have a real-time filter as specified. If the filter is to be part of an off-line model, we only have to ensure that the model time advances by 0.01 seconds between executions.

14.6 Software Filters for Data

As well as providing real-time filtering for control systems, software filters can be used for smoothing or sharpening data.

Suppose that we have an array of sampled results, stored in v[0] to v[200]. We can smooth these by applying the equivalent of a low-pass filter to them:

```
T = 50;
x = v[0];
for (i = 2; i < 201; i++) {
    x = x + (v[i]-x)/T;
    v[i] = x;
}
```

The effect is as though the signal has been subjected to a low-pass filter with time constant equal to 50 times the sample interval. If you decrease the value of *T*, the filter time constant will reduce. If the array is set up to represent a signal with a step change, and the values are plotted before and after filtering, then the result will appear as in Figure 14.5. The values have been "smeared" to the right.

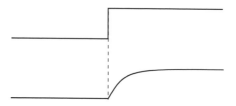

Figure 14.5   Array with step, low-pass filter.

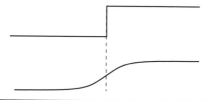

Figure 14.6   A step filtered by $1/(1 + T^2\omega^2)$.

With a stored array of samples we could smear the results to the left instead, by performing the computation in the reverse order. If we apply two filters, one forward and one backward, we can balance the smearing completely:

```
T = 50;
x = v[0];
for (i = 1; i < 201; i ++) {
    x = x + (v[i] -x) /T;
    v[i] = x;
}
for (i = 199; i > -1; i--) {
    x = x + (v[i] -x) /T;
    v[i] = x;
}
```

The result is shown in Figure 14.6. It is equivalent to processing by the second order filter

$$\frac{1}{(1+Ts)(1-Ts)}$$

$$=\frac{1}{1-T^2s^2}$$

If we write $j\omega$ for s, we see the frequency response to be

$$\frac{1}{1+T^2\omega^2}$$

This is always real; the filter has zero phase shift at all frequencies.

To apply such a filter to a real-time signal is of course out of the question. One of the poles is hopelessly unstable, and it is only because we have applied its contribution in "negative time" that we have been able to get away with the trick. Look again at the way the filtered signal starts to change before the step arrives, and you will see that the filter is "non-causal"—the output depends not just on the past values of the input signal, but also on its future.

You can see an image-smoothing application at www.esscont.com/14/smooth. htm.

14.7 State Equations in the Companion Form

If the required filter is causal, and if we have a transfer function to specify it, how can we arrive at a suitable set of state equations to construct it? Suppose that the transfer function is the ratio of two polynomials, $Z(s)/P(s)$, where as before $Z(s)$ defines the zeros while $P(s)$ defines the poles. Take a practical example,

$$G(s) = \frac{4s^2 + s + 1}{s^3 + 3s^2 + 4s + 2}$$

so

$$Z(s) = 4s^2 + s + 1$$

and

$$P(s) = s^3 + 3s^2 + 4s + 2$$

We can build a set of state equations in "companion form" as follows. If we make the second state variable the derivative of the first, and the third the derivative of the second and so on, then we have

$$\dot{x}_1 = x_2$$

$$\dot{x}_2 = x_3$$

where the state variables represent successively higher derivatives of the first.

Now it is clear that if the third state equation is

$$\dot{x}_3 = -3x_3 - 4x_2 - 2x_1 + u$$

it can be rearranged as

$$\dot{x}_3 + 3x_3 + 4x_2 + 2x_1 = u$$

This will be equivalent to

$$\dddot{x}_1 + 3\ddot{x}_1 + 4\dot{x}_1 + 2x_1 = u$$

If we take the Laplace transform, we have

$$(s^3 + 3s^2 + 4s + 2)X_1(s) = U(s)$$

That gives us a system with the correct set of poles. In matrix form, the state equations are:

$$
\begin{bmatrix} \dot{x}_1 \\ \dot{x}_2 \\ \dot{x}_3 \end{bmatrix} =
\begin{bmatrix} 0 & 1 & 0 \\ 0 & 0 & 1 \\ -3 & -4 & -2 \end{bmatrix}
\begin{bmatrix} \dot{x}_1 \\ \dot{x}_2 \\ \dot{x}_3 \end{bmatrix} +
\begin{bmatrix} 0 \\ 0 \\ 1 \end{bmatrix} u
$$

That settles the denominator. How do we arrange the zeros, though? Our output now needs to contain derivatives of x_1,

$$y = 4\ddot{x}_1 + \dot{x}_1 + x_1$$

But we can use our first two state equations to replace this by

$$y = x_1 + x_2 + 4x_3$$

i.e.,

$$y = \begin{bmatrix} 1 & 1 & 4 \end{bmatrix} \mathbf{x}$$

We can only get away with this form $\mathbf{y} = \mathbf{Cx}$ if there are more poles than zeros. If they are equal in number, we must first perform one stage of "long division" of the numerator polynomial by the denominator to split off a $\mathbf{Du}$ term proportional to the input. The remainder of the numerator will then be of a lower order than the denominator and so will fit into the pattern. If there are more zeros than poles, give up.

Now whether it is a simulation or a filter, the system can be generated in terms of a few lines of software. If we were meticulous, we could find a lot of unanswered questions about the stability of the simulation, about the quality of the approximation and about the choice of step length. For now let us turn our attention to the computational techniques of convolution.

Q 14.7.1

We wish to synthesize the filter $s^2/(s^2 + 2s + 1)$ in software. Set up the state equations and write a brief segment of program.

Chapter 15

Time, Frequency, and Convolution

Although the coming sections might seem something of a mathematician's playground, they are extremely useful for getting an understanding of underlying principles of functions of time and the way that dynamic systems affect them. In fact, many of the issues of convolution can be much more easily be explored in terms of discrete time and sampled systems, but first we will take the more traditional approach of infinite impulses and vanishingly small increments of time.

15.1 Delays and the Unit Impulse

We have already looked into the function of time that has a Laplace transform which is just 1. This is the "delta function" $\delta(t)$ when $t = 0$. The unit step has Laplace transform $1/s$, and so we can think of the delta function as its derivative. Before we go on, we must derive an important property of the Laplace transform, the "shift theorem."

If we have a function of time, $x(t)$, and if we pass this signal through a time delay τ, then the output is the same signal that was input τ seconds earlier, $x(t - \tau)$.

The bilateral Laplace transform of this output will be

$$\int_{-\infty}^{\infty} x(t - t)e^{-st}\,dt$$

If we write T for $t - \tau$, then dt will equal dT, and the integral becomes

$$\int_{-\infty}^{\infty} x(T)e^{-s(T+\tau)}dT$$

$$= e^{-s\tau}\int_{-\infty}^{\infty} x(T)e^{-sT}dT$$

$$= e^{-s\tau}X(s)$$

where $X(s)$ is the Laplace transform of $x(t)$. If we delay a signal by time τ, its Laplace transform is simply multiplied by $e^{-s\tau}$.

Since we are considering the bilateral Laplace transform, integrated over all time both positive and negative, we could consider time advances as well. Clearly all signals have to be very small for large negative t, otherwise their contribution to the integral would be enormous when multiplied by the exponential.

We can immediately start to put the shift theorem to use. It tells us that the transform of $\delta(t - \tau)$, the unit impulse shifted to occur at $t = \tau$, is $e^{-s\tau}$. We could of course have worked this out from first principles.

We can regard the delta function as a "sampler." When we multiply it by any function of time, $x(t)$ and integrate over all time, we will just get the contribution from the product at the time the delta function is non-zero.

$$\int_{-\infty}^{\infty} x(t)\delta(t - \tau)dt = x(\tau) \tag{15.1}$$

So when we write

$$\mathcal{L}\big(\delta(t - \tau)\big) = \int_{-\infty}^{\infty} e^{-st}\delta(t - \tau)dt$$

we can think of the answer as sampling e^{-st} at the value $t = \tau$.

Let us briefly indulge in a little philosophy about the "meaning" of functions. We could think of $x(t)$ as a simple number, the result of substituting some value of t into a formula for computing x.

We can instead expand our vision of the function to consider the whole graph of $x(t)$ plotted against time, as in a step response. In control theory we have to take this broader view, regarding inputs and outputs as time "histories," not just as simple values. This is illustrated in Figure 15.1.

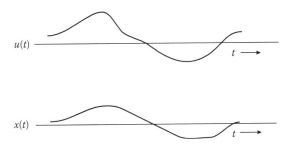

Figure 15.1   Input and output as time plots.

Now we can view Equation 15.1 as a sampling process, allowing us to pick one single value of the function out of the time history. But just let us exchange the symbols t and τ in the equation and suddenly the perspective changes. The substitution has no absolute mathematical effect, but it expresses our time history $x(t)$ as the sum of an infinite number of impulses of size $x(\tau)d\tau$,

$$x(t) = \int_{-\infty}^{\infty} x(\tau)\delta(\tau - t)d\tau \qquad (15.2)$$

This result may not look important, but it opens up a whole new way of looking at the response of a system to an applied input.

15.2 The Convolution Integral

Let us first define the situation. We have a system described by a transfer function $G(s)$, with input function $u(t)$ and output $y(t)$, as in Figure 15.2.

If we apply a unit impulse to the system at $t = 0$, the output will be $g(t)$, where the Laplace transform of $g(t)$ is $G(s)$. This is portrayed in Figure 15.3.

How do we go about deducing the output function for any general $u(t)$?

Perhaps the most fundamental property of a linear system is the "principle of superposition." If we know the output response to a given input function and also to another function, then if we add the two input functions together and apply them, the output will be the sum of the two corresponding output responses.

In mathematical terms, if $u_1(t)$ produces the response $y_1(t)$ and $u_2(t)$ produces response $y_2(t)$, then an input of $u_1(t) + u_2(t)$ will give an output $y_1(t) + y_2(t)$.

Now an input of the impulse $\delta(t)$ to $G(s)$ provokes an output $g(t)$. An impulse applied at time $t = \tau$, $u(\tau)\delta(t - \tau)$ gives the delayed response $u(\tau)g(t - \tau)$. If we apply several impulses in succession, the output will be the sum of the individual responses, as shown in Figure 15.4.

Figure 15.2   Time-functions and the system.

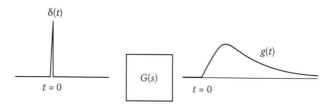

Figure 15.3   For a unit impulse input, *G(s)* gives an output *g(t)*.

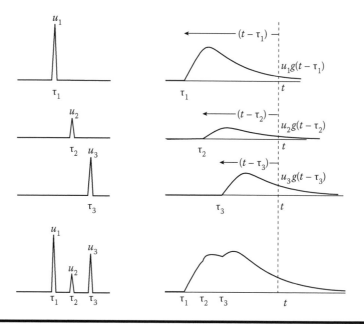

Figure 15.4   Superposition of impulse responses.

Notice that as the time parameter in the *u*-bracket increases, the time in the *g*-bracket reduces. At some later time *t*, the effect of the earliest impulse will have had longest to decay. The latest impulse has an effect that is still fresh.

Now we see the significance of Equation 15.2. It allows us to express the input signal $u(t)$ as an infinite train of impulses $u(\tau)\delta\tau\ \delta(t-\tau)$. So to calculate the output, we add all the responses to these impulses. As we let δτ tend to zero, this becomes the integral,

$$y(t) = \int_{-\infty}^{\infty} u(\tau)g(t-\tau)d\tau \qquad (15.3)$$

This is the convolution integral.

We do not really need to integrate over all infinite time. If the input does not start until $t = 0$ the lower limit can be zero. If the system is "causal," meaning that it cannot start to respond to an input before the input happens, then the upper limit can be t.

15.3 Finite Impulse Response (FIR) Filters

We see that instead of simulating a system to generate a filter's response, we could set up an impulse response time function and produce the same result by convolution. With infinite integrals lurking around the corner, this might not seem such a wise way to proceed!

In looking at digital simulation, we have already cut corners by taking a finite step-length and accepting the resulting approximation. A digital filter must similarly accept limitations in its performance in exchange for simplification. Instead of an infinite train of impulses, $u(t)$ is now viewed as a train of samples at finite intervals. The infinitesimal $u(\tau)d\tau$ has become $u(nT)T$. Instead of impulses, we have numbers to input into a computational process.

The impulse response function $g(t)$ is similarly broken down into a train of sample values, using the same sampling interval. Now the infinitesimal operations of integration are coarsened into the summation

$$y(nT) = \sum_{r=-\infty}^{r=\infty} Tu(rT)g((n-r)T) \qquad (15.4)$$

The infinite limits still do not look very attractive. For a causal system, however, we need go no higher than $r = n$, while if the first signal was applied at $r = 0$ then this can be the lower limit.

Summing from $r = 0$ to n is a definite improvement, but it means that we have to sum an increasing number of terms as time advances. Can we do any better?

Most filters will have a response which eventually decays after the initial impulse is applied. The one-second lag $1/(s+1)$ has an initial response of unity, gives an output of around 0.37 after one second, but after 10 seconds the output has decayed to less than 0.00005. There is a point where $g(t)$ can safely be ignored, where indeed it is less than the resolution of the computation process. Instead of regarding the impulse response as a function of infinite duration, we can cut it short to become a Finite Impulse Response. Why the capital letters? Since this is the basis of the FIR filter.

We can rearrange Equation 15.4 by writing $n - r$ instead of r and vice versa. We get

$$y(nT) = \sum_{r=-\infty}^{r=\infty} Tu((n-r)T)g(rT)$$

Now if we can say that $g(rT)$ is zero for all $r < 0$, and also for all $r > N$, the summation limits become

$$y(nT) = \sum_{r=0}^{r=N} Tu((n-r)T)g(rT)$$

The output now depends on the input u at the time in question, and on its past N values. These values are now multiplied by appropriate fixed coefficients and summed to form the output, and are moved along one place to admit the next input sample value. The method lends itself ideally to a hardware application with a "bucket-brigade" delay line, as shown in Figure 15.5.

The following software suggestion can be made much more efficient in time and storage; it concentrates on showing the method. Assume that the impulse response has already been set up in the array $g(i)$, where i ranges from 0 to N. We provide another array $u(i)$ of the same length to hold past values.

```
//Move up the input samples to make room for a new one
for (i=N; i>0; i--) {
    u[i]=u[i-1];
}
//Take in a new sample
u[0]=GetNewInput();
```

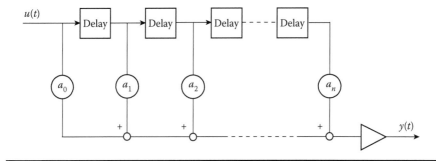

Figure 15.5 A FIR filter can be constructed from a "bucket-brigade" delay line.

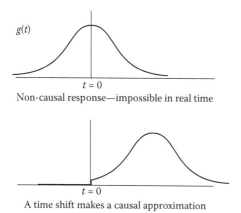

$g(t)$

$t = 0$

Non-causal response—impossible in real time

$t = 0$

A time shift makes a causal approximation

Figure 15.6   By delaying a non-causal response, it can be made causal.

```
//Now compute the output
y=0;
for(i=0;i<N+1;i++){
    y=y+u[i]*g[i];
}

//y now holds the output value
```

This still seems more trouble than the simulation method; what are the advantages? Firstly, there is no question of the process becoming unstable. Extremely sharp filters can be made for frequency selection or rejection which would have poles very close to the stability limit. Since the impulse response is defined exactly, stability is assured.

Next, the rules of causality can be bent a little. Of course the output cannot precede the input, but by considering the output signal to be delayed the impulse response can have a "leading tail." Take the non-causal smoothing filter discussed earlier, for example. This has a bell-shaped impulse response, symmetrical about $t=0$ as shown in Figure 15.6. By delaying this function, all the important terms can be contained in a positive range of t. There are many applications, such as offline sound and picture filtering, where the added delay is no embarrassment.

15.4 Correlation

This is a good place to give a mention to that close relative of convolution, *correlation*. You will have noticed that convolution combines two functions of time by running the time parameter forward in the one and backward in the other. In correlation the parameters run in the same direction.

The use of correlation is to compare two time functions and find how one is influenced by the other. The classic example of correlation is found in the satellite global positioning system (GPS). The satellite transmits a *pseudo random binary sequence* (PRBS) which is picked up by the receiver. Here it is correlated against the various signals that are known to have been transmitted, so that the receiver is able to determine both whether the signal is present and by how much it has been delayed on its journey.

So how does it do it?

The correlation integral is

$$\Phi_{xy}(\tau) = \int x(t)y(t+\tau)dt \tag{15.5}$$

giving a function of the time-shift between the two functions. The *coarse acquisition* signal used in GPS for public, rather than military purposes, is a PRBS sequence of length 1023. It can be regarded as a waveform of levels of value +1 or −1. If we multiply the sequence by itself and integrate over one cycle, the answer is obviously 1023. What makes the sequence "pseudo random" is that if we multiply its values by the sequence shifted by any number of pulses, the integral gives just −1.

Figure 15.7 shows all 1023 pulses of such a sequence as lines of either black or white. The autocorrelation function, the correlation of the sequence with itself, is as shown in Figure 15.8, but here the horizontal scale has been expanded so show

Figure 15.7   Illustration of a pseudo-random sequence.

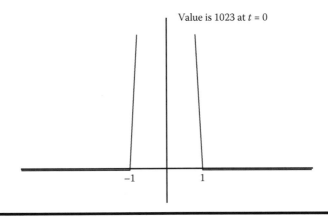

Figure 15.8   Autocorrelation function of a PRBS.

just a few pulse-widths of shift. In fact this autocorrelation function repeats every 1023 shifts; it is cyclic.

When we correlate the transmitted signal against the signal at the receiver, we will have a similar result to Figure 15.8, but shifted by the time it takes for the transmitted signal to reach the receiver. In that way the distance from the satellite can be estimated. (The receiver can reconstruct the transmitted signal because it "knows" the time and the formula that generates the sequence.)

There are in fact 32 different "songs" of this length that the satellites can transmit and the cross-correlation of any pair will be zero. Thus, the receiver can identify distances to any of the satellites that are in view. From four or more such distances, using an *almanac* and *ephemeris* to calculate the exact positions of the satellites, the receiver can solve for *x*, *y*, *z*, and *t*, refining its own clock's value to a nanosecond.

So what is a treatise on GPS doing in a control textbook?

There are some valuable principles to be seen. In this case, the "transfer function" of the path from satellite to receiver is a time-delay. The cross-correlation enables this to be measured accurately. What happens when we calculate the cross-correlation between the input and the output of any control system? We have

$$\Phi_{uy}(\tau) = \int_t u(t)\,y(t+\tau)dt$$

Equation 15.3 tells us that

$$y(t) = \int_{\tau=-\infty}^{\infty} u(\tau)g(t-\tau)d\tau$$

i.e.,

$$y(t) = \int_{T=-\infty}^{\infty} u(T)g(t-T)dT$$

$$y(t+\tau) = \int_{T=-\infty}^{\infty} u(T)g(t+\tau-T)dT$$

So

$$\Phi_{uy}(\tau) = \int_t u(t)\left(\int_T u(T)g(t+\tau-T)dT \right)dt$$

$$\Phi_{uy}(\tau) = \int_t u(t)\left(\int_T u(T+t)g(\tau-T)dT \right)dt$$

Now if we reverse the order of integration, we have

$$\Phi_{uy}(\tau) = \int_T \left(\int_t u(t)u(T+t)dt \right) g(\tau-T)dt$$

$$\Phi_{uy}(\tau) = \int_T \left(\Phi_{uu}(T) \right) g(\tau-T)dt \qquad (15.6)$$

The cross-correlation function is the function of time that would be output from the system if the input's autocorrelation function were applied instead of the input. This illustrated in Figure 15.9.

Provided we have an input function that has enough bandwidth, so that its autocorrelation function is "sharp enough," we can deduce the transfer function by cross-correlation. This can enable adaptive controllers to adapt.

More to the point, we can add a PRBS to any linear input and so "tickle" the system rather than hitting it with a single impulse. In the satellite system, the

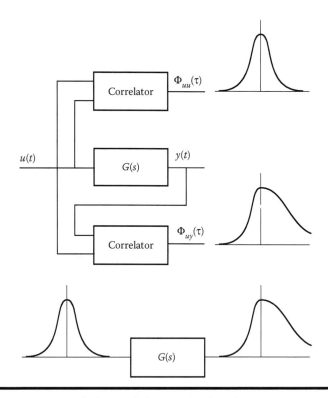

Figure 15.9   Cross-correlation and the transfer function.

PRBS is clocked at one megahertz and repeats after a millisecond. But it is easy to construct longer sequences. One such sequence only repeats after a month! So in a multi-input systems, orthogonal sequences can be applied to various inputs to identify the individual transfer functions as impulse responses.

15.5 Conclusion

We have seen that the state description of a system is bound tightly to its representation as an array of transfer functions. We have requirements which appear to conflict. One the one hand, we seek a formalism which will allow as much of the work as possible to be undertaken by computer. On the other, we wish to retain an insight into the nature of the system and its problems, so that we can use intelligence in devising a solution.

Do we learn more from the time domain, thinking in terms of matrix equations and step and impulse responses, or does the transfer function tell us more, with its possibilities of frequency response and root locus?

In the next chapter, we will start to tear the state equations apart to see what the system is made of. Maybe we can get the best of both worlds.

Chapter 16

More about Time and State Equations

16.1 Introduction

It is time to have a thorough look at the nature of linear systems, the ways in which their state equations can be transformed and the formal analysis of dynamic compensators.

We have now seen the same system described by arrays of transfer functions, differential equations and by first-order matrix state equations. We have seen that however grandiose the system may be, the secret of its behavior is unlocked by finding the roots of a single polynomial characteristic equation. The complicated time solution then usually crumbles into an assortment of exponential functions of time.

In this chapter we are going to hit the matrix state equations with the power of algebra, to open up the can of worms and simplify the structure inside it. We will see that a transformation of variables will let us unravel the system into an assortment of simple subsystems, whose only interaction occurs at the input or the output.

16.2 Juggling the Matrices

We start with the now familiar matrix state equations

$$\dot{\mathbf{x}} = \mathbf{A}\mathbf{x} + \mathbf{B}\mathbf{u}$$

$$\mathbf{y} = \mathbf{C}\mathbf{x}$$

$$(16.1)$$

When we consider a transformation to a new set of state variables, **w**, where **w** and **x** are related by the transformation

$$\mathbf{w} = \mathbf{Tx}$$

with inverse transformation

$$\mathbf{x} = \mathbf{T}^{-1}\mathbf{w}$$

then we find that

$$\dot{\mathbf{w}} = \mathbf{TAT}^{-1}\mathbf{w} + \mathbf{TBu}$$

$$\mathbf{y} = \mathbf{CT}^{-1}\mathbf{w}$$

(16.2)

The new equations still represent the same system, just expressed in a different set of coordinates, so the "essential" properties of **A** must be unchanged by any valid transformation. What are they?

How can we make the transformation tell us more about the nature of **A**? In Section 5.5 we had a foretaste of transformations and saw that in that particular example the new matrix could be made to be diagonal. Is this generally the case?

16.3 Eigenvectors and Eigenvalues Revisited

In Chapter 8 we previewed the nature of eigenvalues and eigenvectors. Perhaps we should consider some examples to clarify them. Let us take as an example the matrix:

$$\mathbf{A} = \begin{bmatrix} 2 & 2 \\ 3 & 1 \end{bmatrix}$$

If we post-multiply **A** by a column vector, we get another column vector, for example,

$$\begin{bmatrix} 2 & 2 \\ 3 & 1 \end{bmatrix}\begin{bmatrix} 1 \\ 0 \end{bmatrix} = \begin{bmatrix} 2 \\ 3 \end{bmatrix}$$

The direction of the new vector is different and its size is changed. Multiplying by **A** takes $(0, 1)'$ to $(2, 1)'$, takes $(1, -1)'$ to $(0, 2)'$, and takes $(1, 1)'$ to $(4, 4)'$.

Wait a minute, though. Is there not something special about this last example? The vector $(4, 4)'$ is in exactly the same direction as the vector $(1, 1)'$, and is multiplied in size by 4. No matter what set of coordinates we choose, the property that there is a vector on which **A** just has the effect of multiplying it by 4 will be unchanged. Such a vector is called an *eigenvector* and the number by which it is multiplied is an *eigenvalue*.

Is this the only vector on which **A** has this sort of effect? We can easily find out. We are looking for a vector **x** such that $\mathbf{Ax} = \lambda\mathbf{x}$, where λ is a mere constant. Now

$$\mathbf{Ax} = \lambda\mathbf{x}$$

i.e.,

$$\mathbf{Ax} = \lambda\mathbf{Ix},$$

where **I** is the unit matrix, so

$$(\mathbf{A} - \lambda\mathbf{I})\mathbf{x} = 0. \tag{16.3}$$

This **0** is a vector of zeros. We have a respectable vector **x** being multiplied by a matrix to give a null vector. Multiplying a matrix by a vector can be regarded as mixing the column vectors that make up the matrix, to arrive at another column vector. Here the components of **x** mix the columns of **A** to get a vector of zeros.

Now one way of simplifying the calculation of a determinant is to add columns together in a mixture that will make more of the coefficents zero. But here we arrive at a column that is all zero, so the determinant must be zero.

$$\det(\mathbf{A} - \lambda\mathbf{I}) = 0. \tag{16.4}$$

Write *s* in place of λ and this should bring back memories. How that characteristic polynomial gets around!

In the example above, the determinant of $\mathbf{A} - \lambda\mathbf{I}$ is

$$\begin{vmatrix} 2-\lambda & 2 \\ 3 & 1-\lambda \end{vmatrix}$$

giving

$$\lambda^2 - 3\lambda - 4 = 0$$

This factorizes into

$$(\lambda + 1)(\lambda - 4) = 0$$

The root $\lambda = 4$ comes as no surprise, and we know that it corresponds to the eigenvector $(1, 1)'$.

Let us find the vector that corresponds to the other value, -1. Substitute this value for λ in Equation 16.3 and we have

$$\begin{bmatrix} 3 & 2 \\ 3 & 2 \end{bmatrix}\begin{bmatrix} x_1 \\ x_2 \end{bmatrix} = \begin{bmatrix} 0 \\ 0 \end{bmatrix}$$

The two simultaneous equations have degenerated into one (if they had not, we could not solve for a single value of x_1/x_2). The vector $(2, -3)'$ is obviously a solution, as is any multiple of it. Multiply it by $\mathbf{A}$, and we are reassured to see that the result is $(-2, 3)$.

In general, if $\mathbf{A}$ is n by n we will have n roots of the characteristic equation, and we should be able to find n eigenvectors to go with them. If the roots, the eigenvalues, are all different, it can be shown that the eigenvectors are all linearly independent, meaning that however we combine them we cannot get a zero vector. We can pack these columns together to make a respectable transformation matrix, $\mathbf{T}^{-1}$. When this is pre-multiplied by $\mathbf{A}$ the result is a pack of columns that are the original eigenvectors, each now multiplied by its eigenvalue.

If we pre-multiply this by the inverse transformation, $\mathbf{T}$, we arrive at the elegant result of a diagonal matrix, each element of which is one of the eigenvalues.

To sum up:

First solve $\det(\mathbf{A} - \lambda\mathbf{I}) = 0$ to find the eigenvalues.

Substitute the n eigenvalues in turn into $(\mathbf{A} - \lambda\mathbf{I})\mathbf{x} = \mathbf{0}$ to find the n eigenvectors in the form of column vectors.

Pack these column vectors together, in order of decreasing eigenvalue for neatness, to make a matrix $\mathbf{T}^{-1}$. Find its inverse, $\mathbf{T}$.

We now see that

$$\mathbf{TAT}^{-1} = \begin{bmatrix} \lambda_1 & 0 & 0 & \cdots \\ 0 & \lambda_2 & 0 & \cdots \\ 0 & 0 & \lambda_3 & \cdots \\ \cdots & \cdots & \cdots & \cdots \end{bmatrix} \tag{16.5}$$

Q 16.3.1

Perform the above operations on the matrix

$$\begin{bmatrix} 0 & 1 \\ -6 & -5 \end{bmatrix}$$

then look back at Section 5.5.

16.4 Splitting a System into Independent Subsystems

As soon as we have a diagonal system matrix, the system falls apart into a set of unconnected subsystems. Each equation for the derivative of one of the w's contains only that same w on the right-hand side,

$$\dot{w}_n = \lambda_n w_n + (\mathbf{TBu})_n \qquad (16.6)$$

In the "companion form" described in Section 14.7, we saw that a train of off-axis 1's expressed each state variable as the derivative of the one before, linking them all together. There is no such linking in the diagonal case. Each variable stands on its own. The only coupling between variables is their possible sharing of the inputs, and their mixture to form the outputs, as shown in Figure 16.1.

Each component of w represents one exponential function of time in the analytic solution of the differential equations. The equations are equivalent to taking each transfer function expression, factorizing its denominator and splitting it into partial fractions.

Let us now suppose that we have made the transformation and we have a diagonal $\mathbf{A}$, with new state equations expressed in variables that we have named $\mathbf{x}$.

When the system has a single input and a single output we can be even more specific about the shape of the system. Such a system might for example be a filter, represented by one single transfer function. Now each of the subsystems is driven from the single input, and the outputs are mixed together to form the single output. The matrix $\mathbf{B}$ is an n-element column vector, while $\mathbf{C}$ is an n-element row vector.

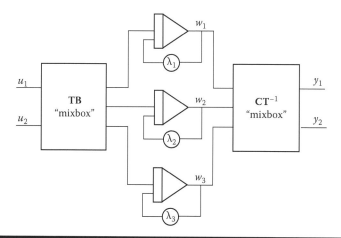

Figure 16.1 The TB matrix provides a mixture of inputs to the independent state integrators.

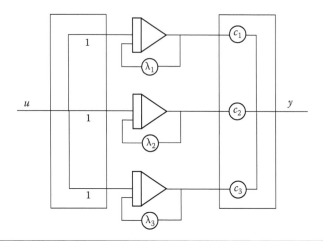

Figure 16.2   A SISO system with unity B matrix coefficients.

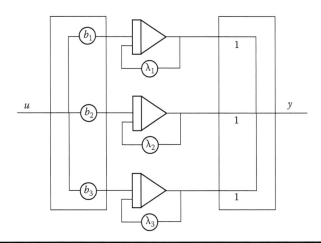

Figure 16.3   A SISO system with unity C matrix coefficients.

If we double the values of all the **B** elements and halve all those of **C**, the overall effect will be exactly the same. In fact the gain associated with each term will be the product of a **B** coefficient and the corresponding **C** coefficient. Only this product matters, so we can choose each of the **B** values to be unity and let **C** sort out the result, as shown in Figure 16.2. Equally we could let the **C** values be unity, and rely on **B** to set the product as shown in Figure 16.3.

In matrix terms, the transformation matrix, $\mathbf{T}^{-1}$ is obtained by stacking together the column vectors of the eigenvectors. Each of these vectors could be multiplied by a constant, and it would still be an eigenvector. The transformation matrix

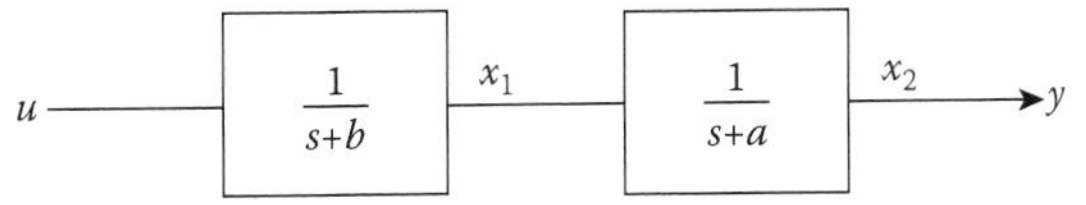

Figure 16.4   Two cascaded lags.

is therefore far from unique, and can be fiddled to make **B** or **C** (in the single-input–single-output case) have elements that are all unity.

Let us bring the discussion back down to earth by looking at an example or two. Consider the second-order lag, or low-pass filter, defined by the transfer function:

$$Y(s) = \frac{1}{(s+a)(s+b)} U(s)$$

The obvious way to split this into first-order subsystems is to regard it as two cascaded lags, as in Figure 16.4.

$$Y(s) = \frac{1}{(s+a)} \frac{1}{(s+b)} U(s)$$

Applying the two lags one after the other suggests a state-variable representation

$$\dot{x}_1 = -bx_1 + u$$

$$\dot{x}_2 = -ax_2 + x_1$$

$$y = x_2$$

i.e.,

$$\dot{\mathbf{x}} = \begin{bmatrix} -b & 0 \\ 1 & -a \end{bmatrix} \mathbf{x} + \begin{bmatrix} 1 \\ 0 \end{bmatrix} u$$

$$y = \begin{bmatrix} 0 & 1 \end{bmatrix} \mathbf{x}$$

Transforming the A-matrix to diagonal form is equivalent to breaking the transfer function into partial fractions:

$$Y(s) = \frac{1}{(b-a)} \left(\frac{1}{(s+a)} - \frac{1}{(s+b)} \right) U(s)$$

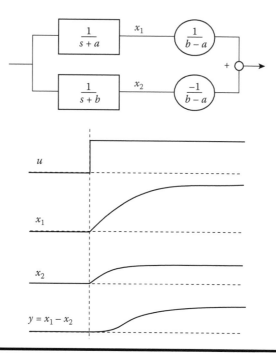

Figure 16.5   **A second-order system as the difference of first-order subsystems.**

so

$$Y(s) = \frac{1}{(b-a)}\left(X_1(s) - X_2(s)\right)$$

where

$$X_1(s) = \frac{1}{s+a}U(s)$$

$$X_2(s) = \frac{1}{s+b}U(s)$$

So we have represented this second-order lag as the difference between two first-order lags, as shown in Figure 16.5.

The second-order lag response to a step input function has zero initial derivative. The two first-order lags are mixed in such a way that their initial slopes are canceled.

Q 16.4.1

Write out the state equations of the above argument in full, for the case where $a = 2$, $b = 3$.

Q 16.4.2

Write out the equations again for the case $a = b = 2$. What goes wrong?

16.5 Repeated Roots

There is always a snag. If all the roots are distinct, then the A-matrix can be made diagonal using a matrix found from the eigenvectors and all is well. A repeated root throws a spanner in the works in the simplest of examples.

In Section 16.4 we saw that a second-order step response could usually be derived from the difference between two first-order responses. But if the time constants coincide the difference becomes zero, and hence useless.

The analytic solution contains not just a term e^{-at} but another term te^{-at}. The exponential is multiplied by time.

We simply have to recognize that the two cascaded lags can no longer be separated, but must be simulated in that same form. If there are three equal roots, then there may have to be three equal lags in cascade. Instead of achieving a diagonal form, we may only be able to reduce the A-matrix to a form such as

$$
A = \left[\begin{array}{ccc|ccc}
a & 1 & 0 & 0 & 0 & 0 \\
0 & a & 1 & 0 & 0 & 0 \\
0 & 0 & a & 0 & 0 & 0 \\
\hline
0 & 0 & 0 & b & 1 & 0 \\
0 & 0 & 0 & 0 & b & 0 \\
0 & 0 & 0 & 0 & 0 & c
\end{array}\right]
\tag{16.7}
$$

This is the *Jordan Canonical Form*, illustrated in Figure 16.6.

Repeated roots do not always mean that a diagonal form is impossible. Two completely separate single lags, each with their own input and output, can be combined into a singe set of system equations. The A-matrix is then of course diagonal, since there is no reaction between the two subsystems. If the system is single-input–single-output, however, repeated roots always mean trouble.

In another case, the diagonal form is possible but not the most desirable. Suppose that some of the roots of the characteristic equation are complex. It is not easy to apply a feedback gain of $2 + 3j$ around an integrator!

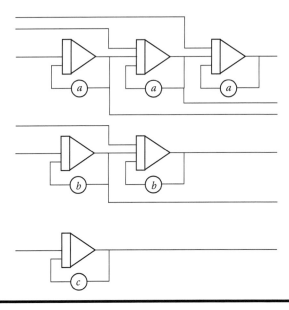

Figure 16.6 Simulation structure of a system with repeated roots.

The second-order equation with roots $-k + / -j \cdot n$,

$$\ddot{y} + 2k\dot{y} + (k^2 + n^2)y = u$$

is more neatly represented for simulation by

$$\begin{bmatrix} \dot{x}_1 \\ \dot{x}_2 \end{bmatrix} = \begin{bmatrix} -k & n \\ -n & -k \end{bmatrix} \begin{bmatrix} x_1 \\ x_2 \end{bmatrix} + \begin{bmatrix} 0 \\ 1 \end{bmatrix} u \qquad (16.8)$$

than by a set of matrices with complex coefficients. If we accept quadratic terms as well as linear factors, any polynomial with real coefficients can be factorized without having to resort to complex numbers.

Q 16.5.1

Derive state equations for the system $\ddot{y} + y = u$. Find the Jordan Canonical Form, and also find a form in which a real simulation is possible, similar to the example of expression 16.8. Sketch the simulation.

16.6 Controllability and Observability

With the system equation reduced to diagonal form, we saw the state variables standing alone, with no cross coupling between them. The only signals that can affect their behavior are the inputs. Suppose, however, that one or more of these state variables has no input connected to it. Then there is no way in which those variables can be controlled. Such variables are "uncontrollable."

If the poles associated with all the uncontrollable variables are stable, then the system performance as a whole can still be acceptable. No amount of feedback can move the positions of these poles, however.

Another alarming possibility is that there are state variables that affect none of the outputs. Their behavior is not measured in any way and so feedback will again have no effect on their pole values. These variables are "unobservable." You may say that since the variables do not affect the outputs, you do not care what they do. Unfortunately the outputs for control purposes are the transducers and sensors of the control system. There might be additional state variables that are in need of control, but for which there are no sensors.

Some of the variables could be both uncontrollable and unobservable, that is to say, they might have neither input nor output connections. The system can now be partitioned into four subsystems, as illustrated in Figure 16.7:

1. Controllable and observable
2. Controllable but not observable

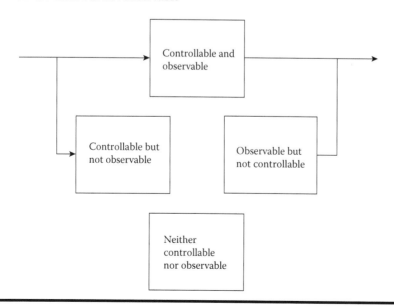

Figure 16.7   Four partitions of a system.

3. Observable but not controllable
4. Neither controllable not observable

Clearly any system model built to represent a transfer function must be both controllable and observable—there is no point in adding uncontrollable or unobservable modes. The control engineer is baffled by the fourth type of subset, somewhat frustrated by sets two and three, and is really only at home in applying control to the first.

Note that a system can be controllable and observable with very few inputs and outputs. In the two cascaded lags example of Section 16.4, we saw that the single input was applied to the first lag, while the single output was taken from the second. So how can we be sure that we can control the state corresponding to the second lag or observe that of the first? When the system is represented in diagonal form all the necessary input and output coefficients are non-zero, ensuring controllability and observability. Can we deduce these properties form the general matrix state equations, without going to the trouble of transforming them?

Before discussing this further, we must define what we mean by the *rank* of a matrix.

When considering transformations, we have already viewed a matrix as a pack of column vectors. Without much definition we have specified that these vectors must be *linearly independent*. Now if we consider the matrix

$$\begin{bmatrix} 1 & 2 \\ 1 & 2 \end{bmatrix}$$

its rank is only 1, even though it is a 2 by 2 matrix. The second column is a multiple of the first column, so if when consider that multiplying the matrix by a vector is the same as taking a linear combination of the columns, we see that the resulting vector can only be in one single direction where in this case the components are equal.

Similarly

$$\begin{bmatrix} 1 & 1 & 2 \\ 2 & 2 & 4 \\ 3 & 4 & 7 \end{bmatrix}$$

has a rank of just two, since although the first two columns are linearly independent, the third is a combination of the first two.

The determinant will only be non-zero if all the columns are linearly independent.

But now the concept of rank gets more interesting. What is the rank of

$$\begin{bmatrix} 1 & 0 & 0 & 2 \\ 0 & 1 & 0 & 3 \\ 0 & 0 & 1 & 4 \end{bmatrix}$$

The first three columns are clearly independent, but the fourth cannot add to the rank. The rank is three. So what has this to do with controllability?

Suppose that our system is of third order. Then by manipulating the inputs, we must be able to make the state vary through all three dimensions of its variables.

$$\dot{\mathbf{x}} = \mathbf{A}\mathbf{x} + \mathbf{B}\mathbf{u}$$

If the system has a single input, then the B matrix has a single column. When we first start to apply an input we will set the state changing in the direction of that single column vector of **B**. If there are two inputs then **B** has two columns. If they are independent the state can be sent in the direction of any combination of those vectors. We can stack the columns of **B** side by side and examine their rank.

But what happens next? When the state has been sent off in such a direction, the **A** matrix will come into play to move the state on further. The velocity can now be in a direction of any of the columns of **AB**. But there is more. The **A** matrix can send each of these new displacements in the direction $\mathbf{A}^2\mathbf{B}$, then $\mathbf{A}^3\mathbf{B}$ and so on forever.

So to test for controllability we look at the rank of the composite matrix

$$\begin{bmatrix} \mathbf{B} & | & \mathbf{A}\mathbf{B} & | & \mathbf{A}^2\mathbf{B} & | & \mathbf{A}^3\mathbf{B} & | & \cdots \end{bmatrix}$$

It can be shown that in the third order case, only the first three terms need be considered. The "Cayley Hamilton" theory shows that if the second power of **A** has not succeeded in turning the state to cover any missing directions, then no higher powers can do any better.

In general the rank must be equal to the order of the system and we must consider terms up to $\mathbf{A}^{n-1}\mathbf{B}$.

It is time for an example.

Q 16.6.1

Is the following system controllable?

$$\dot{\mathbf{x}} = \begin{bmatrix} 0 & 1 \\ -1 & -2 \end{bmatrix} \mathbf{x} + \begin{bmatrix} 0 \\ 1 \end{bmatrix} u$$

Now

$$\mathbf{AB} = \begin{bmatrix} 0 & 1 \\ -1 & -2 \end{bmatrix} \begin{bmatrix} 0 \\ 1 \end{bmatrix} = \begin{bmatrix} 1 \\ -2 \end{bmatrix}$$

so

$$\begin{bmatrix} \mathbf{B} & | & \mathbf{AB} \end{bmatrix} = \begin{bmatrix} 0 & 1 \\ 1 & -2 \end{bmatrix}$$

which has rank 2. The system is controllable.

Q 16.6.2

Is this system also controllable?

$$\dot{\mathbf{x}} = \begin{bmatrix} 0 & 1 \\ -1 & -2 \end{bmatrix} \mathbf{x} + \begin{bmatrix} -1 \\ 1 \end{bmatrix} u$$

Q 16.6.3

Is this third system controllable?

$$\dot{\mathbf{x}} = \begin{bmatrix} 0 & 1 \\ -1 & -2 \end{bmatrix} \mathbf{x} + \begin{bmatrix} 1 \\ 0 \end{bmatrix} u$$

Q 16.6.4

Express that third system in Jordan Canonical Form. What is the simulation structure of that system?

Can we deduce observability in a similar simple way, by looking at the rank of a set of matrices?

The equation

$$\mathbf{y} = \mathbf{C}\mathbf{x}$$

is likely to represent fewer outputs than states; we have not enough simultaneous equations to solve for **x**. If we ignore all problems of noise and continuity, we can consider differentiating the outputs to obtain

$$\dot{y} = C\dot{x}$$

$$= C(Ax + Bu)$$

so

$$\dot{y} - CBu$$

can provide some more equations for solving for *x*. Further differentiation will give more equations, and whether these are enough can be seen by examining the rows of **C**, of **CA** and so on, or more formally by testing the rank of

$$\begin{bmatrix} C \\ \hline CA \\ \hline CA^2 \\ \hline \cdots \\ \hline CA^{n-1} \end{bmatrix}$$

and ensuring that it is the same as the rank of the system.

Showing that the output can reveal the state is just the start. We now have to find ways to perform the calculation without resorting to differentiation.

Chapter 17

Practical Observers, Feedback with Dynamics

17.1 Introduction

In the last chapter, we investigated whether the inputs and outputs of a system made it possible to control all the state variables and deduce their values. Though the tests looked at the possibility of observing the states, they did not give very much guidance on how to go about it.

It is unwise, to say the least, to try to differentiate a signal. Some devices that claim to be differentiators are in fact mere high-pass filters. A true differentiator would have to have a gain that tends to infinity with increasing frequency. Any noise in the signal would cause immense problems.

Let us forget about these problems of differentiation, and instead address the direct problem of deducing the state of a system from its outputs.

17.2 The Kalman Filter

First we have to assume that we have a complete knowledge of the state equations. Can we not then set up a simulation of the system, and by applying the same inputs simply measure the states of the model? This might succeed if the system has only poles that represent rapid settling—the sort of system that does not really need feedback control!

Suppose instead that the system is a motor-driven position controller. The output involves integrals of the input signal. Any error in setting up the initial conditions of the model will persist indefinitely.

Let us not give up the idea of a model. Suppose that we have a measurement of the system's position, but not of the velocity that we need for damping. The position is enough to satisfy the condition for observability, but we do not wish to differentiate it. Can we not use this signal to "pull the model into line" with the state of the system? This is the principle underlying the "Kalman Filter."

The system, as usual, is given by

$$\dot{\mathbf{x}} = \mathbf{A}\mathbf{x} + \mathbf{B}\mathbf{u}$$
$$\mathbf{y} = \mathbf{C}\mathbf{x}$$

(17.1)

We can set up a simulation of the system, having similar equations, but where the variables are $\hat{\mathbf{x}}$ and $\hat{\mathbf{y}}$. The "hats" mark the variables as estimates. Since we know the value of the input, $\mathbf{u}$, we can use this in the model.

Now in the real system, we can only influence the variables through the input via the matrix $\mathbf{B}$. In the model, we can "cheat" and apply an input signal directly to any integrator and hence to any state variable that we choose. We can, for instance, calculate the error between the measured outputs, $\mathbf{y}$, and the estimated outputs $\hat{\mathbf{y}}$ given by $\mathbf{C}\hat{\mathbf{x}}$, and mix this signal among the state integrators in any way we wish. The model equations then become

$$\dot{\hat{\mathbf{x}}} = \mathbf{A}\hat{\mathbf{x}} + \mathbf{B}\mathbf{u} + \mathbf{K}(\mathbf{y} - \mathbf{C}\hat{\mathbf{x}})$$
$$\hat{\mathbf{y}} = \mathbf{C}\hat{\mathbf{x}}$$

(17.2)

The corresponding system is illustrated in Figure 17.1.

The model states now have two sets of inputs, one corresponding to the plant's input and the other taken from the system's measured output. The model's A-matrix has also been changed, as we can see by rewriting 17.2 to obtain

$$\dot{\hat{\mathbf{x}}} = (\mathbf{A} - \mathbf{K}\mathbf{C})\hat{\mathbf{x}} + \mathbf{B}\mathbf{u} + \mathbf{K}\mathbf{y}$$

(17.3)

To see just how well we might succeed in tracking the system state variables, we can combine Equations 17.1 through 17.3 to give a set of differential equations for the estimation error, $\mathbf{x} - \hat{\mathbf{x}}$:

$$\frac{d}{dt}(\mathbf{x} - \hat{\mathbf{x}}) = (\mathbf{A} - \mathbf{K}\mathbf{C})(\mathbf{x} - \hat{\mathbf{x}})$$

(17.4)

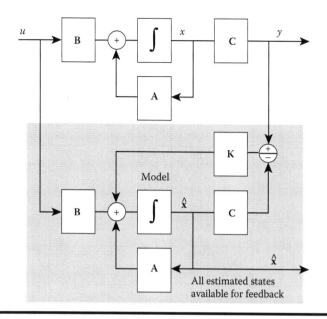

Figure 17.1   Structure of Kalman Filter.

The eigenvalues of $(\mathbf{A} - \mathbf{KC})$ will determine how rapidly the model states settle down to mimic the states of the plant. These are the roots of the model, as defined in Equation 17.3. If the system is observable, we should be able to choose the coefficients of K to place the roots wherever we wish; the choice will be influenced by the noise levels we expect to find on the signals.

Q 17.2.1

A motor is described by two integrations, from input drive to output position. The velocity is not directly measured. We wish to achieve a well-damped position control, and so need a velocity term to add. Design a Kalman observer.

The system equations for this example may be written

$$\dot{\mathbf{x}} = \begin{bmatrix} 0 & 1 \\ 0 & 0 \end{bmatrix} \mathbf{x} + \begin{bmatrix} 0 \\ 1 \end{bmatrix} u$$

$$\mathbf{y} = \begin{bmatrix} 1 & 0 \end{bmatrix} \mathbf{x}$$

(17.5)

The Kalman feedback matrix will be a vector $(p, q)'$ so the structure of the filter will be as shown in Figure 17.2.

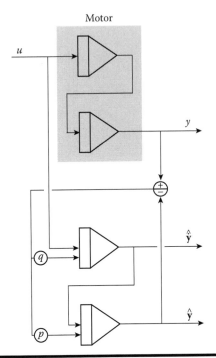

Figure 17.2 Kalman filter to observe motor velocity.

For the model, we have

$$A - KC = \begin{bmatrix} 0 & 1 \\ 0 & 0 \end{bmatrix} - \begin{bmatrix} p \\ q \end{bmatrix} \begin{bmatrix} 1 & 0 \end{bmatrix}$$

$$= \begin{bmatrix} -p & 1 \\ -q & 0 \end{bmatrix}$$

The eigenvalues are given by the determinant $\det(\mathbf{A} - \mathbf{KC} - \lambda\mathbf{I})$, i.e.,

$$\lambda^2 + p\lambda + q = 0.$$

We can choose the parameters of **K** to put the roots wherever we like.

It looks as though we now have both position and velocity signals to feed around our motor system. If this is really the case, then we can put the closed loop poles wherever we wish. It seems that anything is possible—keep hoping.

Q 17.2.2

Design observer and feedback for the motor of Equation 17.5 to give a response characteristic having two equal roots of 0.1 seconds, with an observer error characteristic having equal roots of 0.5 seconds. Sketch the corresponding feedback circuit containing integrators.

17.3 Reduced-State Observers

In the last example, it appeared that we have to use a second-order observer to deduce a single state variable. Is there a more economic way to make an observer? Luenberger suggested the answer.

Suppose that we do not wish to estimate all the components of the state, but only a selection given by $\mathbf{Sx}$. We would like to set up a modeling system having states $\mathbf{z}$, while it takes inputs $\mathbf{u}$ and $\mathbf{y}$. The values of the $\mathbf{z}$ components must tend to the signals we wish to observe. This appears less complicated when written algebraically! The observer equations are

$$\dot{\mathbf{z}} = \mathbf{Pz} + \mathbf{Qu} + \mathbf{Ry} \tag{17.6}$$

and we want all the components of $(\mathbf{z}-\mathbf{Sx})$ to tend to zero in some satisfactory way.

Now we see that the derivative of $(\mathbf{z}-\mathbf{Sx})$ is given by

$$\dot{\mathbf{z}} - \mathbf{S}\dot{\mathbf{x}} = \mathbf{Pz} + \mathbf{Qu} + \mathbf{Ry} - \mathbf{S}(\mathbf{Ax} + \mathbf{Bu})$$

$$= \mathbf{Pz} + \mathbf{Qu} + \mathbf{RCx} - \mathbf{SAx} - \mathbf{SBu}$$

$$= \mathbf{Pz} + (\mathbf{RC} - \mathbf{SA})\mathbf{x} + (\mathbf{Q} - \mathbf{SB})\mathbf{u}$$

We would like this to reduce to

$$\dot{\mathbf{z}} - \mathbf{S}\dot{\mathbf{x}} = \mathbf{P}(\mathbf{z} - \mathbf{Sx})$$

where $\mathbf{P}$ represents a system matrix giving rapid settling. For this to be the case,

$$-\mathbf{PS} = (\mathbf{RC} - \mathbf{SA})$$

and

$$\mathbf{Q} - \mathbf{SB} = 0$$

i.e., when we have decided on **S** that determines the variables to observe, we have

$$\mathbf{Q} = \mathbf{SB},\tag{17.7}$$

$$\mathbf{RC} = \mathbf{SA} - \mathbf{PS}.\tag{17.8}$$

Q 17.3.1

Design a first-order observer for the velocity in the motor problem of Equation 17.5.

Q 17.3.2

Apply the observed velocity to achieve closed loop control as specified in problem Q 17.2.2.

The first problem hits an unexpected snag, as you will see.

If we refer back to the system state equations of 17.5, we see that

$$\dot{\mathbf{x}} = \begin{bmatrix} 0 & 1 \\ 0 & 0 \end{bmatrix} \mathbf{x} + \begin{bmatrix} 0 \\ 1 \end{bmatrix} u$$

$$\mathbf{y} = \begin{bmatrix} 1 & 0 \end{bmatrix} \mathbf{x}$$

If we are only interested in estimating the velocity, then we have

$$\mathbf{S} = [0,1].$$

Now

$$\mathbf{Q} = \mathbf{SB} = \begin{bmatrix} 0 & 1 \end{bmatrix} \begin{bmatrix} 0 \\ 1 \end{bmatrix} = 1$$

P becomes a single parameter defining the speed with which z settles, and we may set it to a value of $-k$.

$$\mathbf{SA} - \mathbf{PS} = \begin{bmatrix} 0 & 1 \end{bmatrix} \begin{bmatrix} 0 & 1 \\ 0 & 0 \end{bmatrix} - \begin{bmatrix} -k \end{bmatrix} \begin{bmatrix} 0 & 1 \end{bmatrix}$$

$$= \begin{bmatrix} 0 & 0 \end{bmatrix} + \begin{bmatrix} 0 & k \end{bmatrix}$$

We now need **R**, which in this case will be a simple constant r, such that

$$\mathbf{RC} = \mathbf{SA} - \mathbf{PS}$$

i.e.,

$$r[1,0] = [0,k]$$

and there is the problem! Clearly there is no possible combination to make the equation balance except if r and k are both zero.

Do not give up! We actually need the velocity so that we can add it to a position term for feedback. So suppose that instead of pure velocity, we try for a mixture of position plus a times the velocity. Then

$$\mathbf{Sx} = [1, a]\mathbf{x}.$$

For **Q** we have the equation

$$\mathbf{Q} = \mathbf{SB} = \begin{bmatrix} 1 & a \end{bmatrix} \begin{bmatrix} 0 \\ 1 \end{bmatrix} = a$$

This time

$$\begin{aligned}
\mathbf{SA} - \mathbf{PS} &= \begin{bmatrix} 1 & a \end{bmatrix} \begin{bmatrix} 0 & 1 \\ 0 & 0 \end{bmatrix} - \begin{bmatrix} -k \end{bmatrix} \begin{bmatrix} 1 & a \end{bmatrix} \\
&= \begin{bmatrix} 0 & 1 \end{bmatrix} + \begin{bmatrix} k & k\,a \end{bmatrix} \\
&= \begin{bmatrix} k & 1 + ka \end{bmatrix}
\end{aligned}$$

Now we can equate this to **RC** if

$$r = k$$

and

$$(1 + ka) = 0.$$

From this last equation we see that a must be negative, with value $-1/k$. That would be the wrong sign if we wished to use this mixture alone as feedback.

However, we can subtract z from the position signal to get a mixture that has the right sign.

$$z = x + a\hat{\dot{x}}$$

$$= x - (1/k)\hat{\dot{x}}$$

so the estimated velocity will be given by

$$\hat{\dot{x}} = k(x - z)$$

where

$$\dot{z} = \mathbf{P}z + \mathbf{Q}u + \mathbf{R}y$$

$$= -kz - (1/k)u + kx$$

Mission accomplished.

To see it in action, let us attack problem Q 17.3.2. We require a response similar to that specified in Q 17.2.2. We want the closed loop response to have equal roots of 0.1 seconds, now with a single observer settling time constant of 0.5 seconds.

For the 0.5 second observer time constant, we make $k = 2$.

For the feedback we now have all the states (or their estimates) available, so instead of considering $(\mathbf{A} + \mathbf{BFC})$ we only have $(\mathbf{A} + \mathbf{BF})$ to worry about. To achieve the required closed loop response we can propose algebraic values for the feedback parameters in $\mathbf{F}$, substitute them to obtain $(\mathbf{A} + \mathbf{BF})$ and then from the eigenvalue determinant derive the characteristic equation. Finally we equate coefficients between the characteristic equation and the equation with the roots we are trying to fiddle. Sounds complicated?

In this simple example we can look at its system equations in the "traditional form"

$$\ddot{x} = u.$$

The two time constants of 0.1 seconds imply root values of 10, so when the loop is closed the behavior of x must be described by

$$\ddot{x} + 20\dot{x} + 100x = 0$$

so

$$u = -100x - 20\dot{x}$$

Instead of he velocity we must use the estimated velocity, so we must set the system input to

$$u = -100x - 20\hat{x}$$

$$u = -100x - 20 \times 2(x - z)$$

$$u = -140x + 40z$$

where z is given by the subsystem we have added

$$\dot{z} = -2z - 0.5u + 2y$$

where the output signal y is the position x.

Before leaving the problem, let us look at the observer in transfer function terms. We can write

$$(s + 2)Z = - 0.5U + 2Y$$

so the equation for the input u becomes

$$U = -140Y + 40Z$$

$$= -140Y + 40\frac{-0.5U + 2Y}{s + 2}$$

So, tidying up and multiplying by $(s + 2)$ we have

$$(s + 2)U + 20U = -140(s + 2)Y + 80Y$$

$$(s + 22)U = -(140s + 200)Y$$

$$U = -\frac{140s + 200}{s + 22}Y$$

The whole process whereby we calculate the feedback input u from the position output y has boiled down to nothing more than a simple phase advance! To make matters worse, it is not even a very practical phase advance, since it requires the high-frequency gain to be over 14 times the gain at low frequency. We can certainly expect noise problems from it.

Do not take this as an attempt to belittle the observer method. Over the years, engineers have developed intuitive techniques to deal with common problems, and only those techniques which were successful have survived. The fact that phase advance can be shown to be equivalent to the application of an observer detracts from neither method—just the reverse.

By tackling a problem systematically, analyzing general linear feedback of states and estimated states, the whole spectrum of solutions can be surveyed to make a choice. The reason that the solution to the last problem was impractical had nothing to do with the method, but depended entirely on our arbitrary choice of settling times for the observer and for the closed loop system. We should instead have left the observer time constant as a parameter for later choice, and we could then have imposed some final limitation on the ratio of gains of the resulting phase advance. Through this method we would at least know the implications of each choice.

If the analytic method does have a drawback, it is that it is too powerful. The engineer is presented with a wealth of possible solutions, and is left agonizing over the new problem of how to limit his choice to a single answer. Some design methods are tailored to reduce these choices. As often as not, they throw the baby out with the bathwater.

Let us go on to examine linear control in its most general terms.

17.4 Control with Added Dynamics

We can scatter dynamic filters all around the control system, as shown in Figure 17.3.

The signals shown in Figure 17.3 can be vectors with many components. The blocks represent transfer function matrices, not just simple transfer functions. This means that we have to use caution when applying the methods of *block diagram manipulation*.

In the case of scalar transfer functions, we can unravel complicated structures by the relationships illustrated in Figure 17.4.

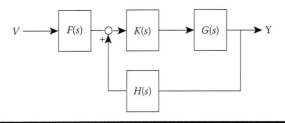

Figure 17.3 Feedback around $G(s)$ with three added filters.

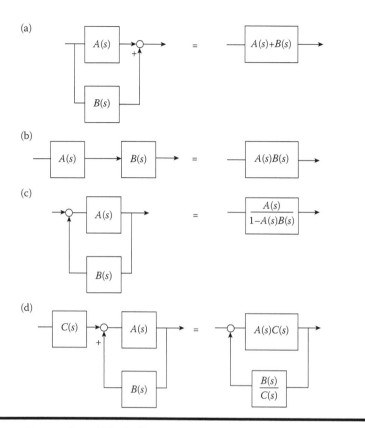

Figure 17.4   Some rules of block diagram manipulation.

Now, however complicated our controller may be, it simply takes inputs from the *command input* **v** and the system output **y** and delivers a signal to the system input **u**. It can be described by just two transfer function matrices. We have one transfer function matrix that links the command input to the system input, which we will call the *feedforward matrix*. We have another matrix linking the output back to the system input, not surprisingly called the *feedback matrix*.

If the controller has internal state **z**, then we can append these components to the system state **x** and write down a set of state equations for our new, bigger system.

Suppose that we started with

$$\dot{\mathbf{x}} = \mathbf{A}\mathbf{x} + \mathbf{B}\mathbf{u}$$
$$\mathbf{y} = \mathbf{C}\mathbf{x}$$

and added a controller with state **z**, where

$$\dot{\mathbf{z}} = \mathbf{Kz} + \mathbf{Ly} + \mathbf{Mv}$$

We apply signals from the controller dynamics, the command input and the system output to the system input, **u**,

$$\mathbf{u} = \mathbf{Fy} + \mathbf{Gz} + \mathbf{Hv}$$

so that

$$\dot{\mathbf{x}} = (\mathbf{A} + \mathbf{BFC})\mathbf{x} + \mathbf{BGz} + \mathbf{BHv}$$

and

$$\dot{\mathbf{z}} = \mathbf{LCx} + \mathbf{Kz} + \mathbf{Mv}$$

We end up with a composite matrix state equation

$$\begin{bmatrix} \dot{\mathbf{x}} \\ \dot{\mathbf{z}} \end{bmatrix} = \left[\begin{array}{c|c} \mathbf{A} + \mathbf{BFC} & \mathbf{BG} \\ \hline \mathbf{LC} & \mathbf{K} \end{array} \right] \begin{bmatrix} \mathbf{x} \\ \mathbf{z} \end{bmatrix} + \begin{bmatrix} \mathbf{BH} \\ \mathbf{M} \end{bmatrix} \mathbf{v}$$

We could even consider a new "super output" by mixing **y, z,** and **v** together, but with the coefficients of all the matrices **F, G, H, K, L,** and **M** to choose, life is difficult enough as it is.

Moving back to the transfer function form of the controller, is it possible or even sensible to try feedforward control alone? Indeed it is.

Suppose that the system is a simple lag, slow to respond to a change of demand. It makes sense to apply a large initial change of input, to get the output moving, and then to turn the input back to some steady value.

Suppose that we have a simple lag with time constant five seconds, described by transfer function

$$\frac{1}{1 + 5s}$$

If we apply a feedforward filter at the input, having transfer function

$$\frac{1 + 5s}{1 + s}$$

(here is that phase advance again!), then the overall response will have the form:

$$\frac{1}{1+s}$$

The five-second time constant has been reduced to one second. We have can-celed a pole of the system by putting an equal zero in the controller, a technique called *pole cancellation*. Figure 17.5 shows the effect.

Take care. Although the pole has been removed from the response, it is still present in the system. An initial transient will decay with a five-second time con-stant, not the one second of our new transfer function. Moreover, that pole has been made uncontrollable in the new arrangement.

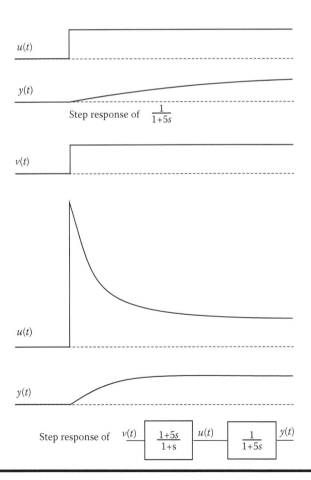

Figure 17.5 Responses with feedforward pole cancellation.

If it is benign, representing a transient that dies away in good time, then we can bid it farewell. If it is close to instability, however, then any error in the controller parameters or any unaccounted "sneak" input or disturbance can lead to disaster.

If we put on our observer-tinted spectacles, we can even represent pole-canceling feedforward in terms of observed state variables. The observer in this case is rather strange in that z can receive no input from y, otherwise the controller would include feedback. Try the example:

Q 17.4.1

Set up the state equations for a five-second lag. Add an estimator to estimate the state from the input alone. Feed back the estimated state to obtain a response with a one-second time constant.

17.5 Conclusion

Despite all the sweeping generalities, the subject of applying control to a system is far from closed. All these discussions have assumed that both system and controller will be linear, when we saw from the start that considerations of signal and drive limiting can be essential.

Chapter 18

Digital Control in More Detail

18.1 Introduction

The philosophers of Ancient Greece found the concept of continuous time quite baffling, getting tied up in paradoxes such as that of Xeno. Today's students are brought up on a diet of integral and differential calculus, taking velocities and accelerations in their stride, so that it is finite differences which may appear quite alien to them.

But when a few fundamental definitions have been established, it appears clear that discrete time systems can be managed more expediently than continuous ones. It is when the two are mixed that the control engineer must be wary.

Even so, we will creep up on the discrete time theory from a basis of continuous time and differential equations.

18.2 Finite Differences—The Beta-Operator

We built our concepts of continuous control on the differential operator, "rate-of-change." We let time change by a small amount δt, resulting in a change of $x(t)$ given by

$$\delta x = x(t + \delta t) - x(t)$$

then we examined the limit of the ratio $\delta x/\delta t$ as we made δt tend to zero.

Now we have let a computer get in on the act, and the rules must change.

$x(t)$ is measured not continuously, but at some sequence of times with fixed intervals. We might know $x(0)$, $x(0.01)$, $x(0.02)$, and so on, but between these values the function is a mystery. The idea of letting δt tend to zero is useless. The only sensible value for δt is equal to the sampling interval. It makes little sense to go on labeling the functions with their time value, $x(t)$. We might as well acknowledge them to be a sequence of discrete samples, and label them $x(n)$ according to their sample number.

We must be satisfied with an approximate sort of derivative, where δt takes a fixed value of one sampling interval, τ. We are left with another subtle problem which is important nonetheless. Should we take our difference as "next value minus this one," or as "this value minus last one?" If we could attribute the difference to lie at a time midway between the samples there would be no such question, but that does not help when we have to tag variables with "sample number" rather than time.

The point of the system equations, continuous or discrete time, is to be able to predict the future state from the present state and the input. If we have the equivalent of an integrator that is integrating an input function $g(n)$, we might write

$$f(n+1) = f(n) + \tau g(n), \tag{18.1}$$

which settles the question in favor of the forward difference

$$g(n) = \{f(n+1) - f(n)\}/\tau.$$

It looks as though we can only calculate $g(n)$ from the $f(n)$ sequence if we already know a future value, $f(n + 1)$. In the continuous case, however, we always dealt with integrators and would not consider differentiating a signal, so perhaps this will not prove a disadvantage.

We might define this approximation to differentiation as an operator, β, where inside it is a "time advance" operation that looks ahead:

$$\beta(f(n)) = \frac{f(n+1) - f(n)}{\tau} \tag{18.2}$$

We could now write

$$g(n) = \beta(f(n)),$$

where we might in the continuous case have used a differential operator to write

$$g(t) = \frac{d}{dt} f(t).$$

The inverse of the differentiator is the integrator, at the heart of all continuous simulation.

The inverse of the β-operator is the crude numerical process of Euler integration as in Equation 18.1.

Now the second order differential equation

$$\ddot{x} + a\dot{x} + bx = u$$

might be approximated in discrete time by

$$\beta^2 x + a\beta x + bx = u$$

Will we look for stability in the same way as before? You will remember that we looked at the roots of the quadratic

$$m^2 + am + b = 0$$

and were concerned that all the roots should have negative real parts. This concern was originally derived from the fact that e^{mt} was an "eigenfunction" of a linear differential equation, that if an input e^{mt} was applied, the output would be proportional to the same function. Is this same property true of the finite difference operator β?

We can calculate

$$\beta(e^{m\tau}) = \frac{e^{(m+1)\tau} - e^{m\tau}}{\tau}$$

$$= \frac{e^\tau - 1}{\tau} e^{m\tau}$$

This is certainly proportional to $e^{m\tau}$. We would therefore be wise to look for the roots of the characteristic equation, just as before, and plot them in the complex plane.

Just a minute, though. Will the imaginary axis still be the boundary between stable and unstable systems? In the continuous case, we found that that a sinewave, $e^{j\omega t}$, emerged from differentiation with an extra multiplying factor $j\omega$, represented by a point on the imaginary axis. Now if we subject the sinewave to the β-operator we get

$$\beta(e^{j\omega t}) = \frac{e^{j\omega \tau} - 1}{\tau} e^{j\omega t}$$

As ω varies, the gain moves along the locus defined by

$$g_r = \text{real part} = \frac{\cos(\omega \tau) - 1}{\tau},$$

$$g_i = \text{imaginary part} = \frac{\sin(\omega \tau)}{\tau}.$$

So

$$(\tau g_r + 1)^2 + \tau g_i^2 = 1.$$

This is a circle, center $-1/\tau$, radius $1/\tau$, which touches the imaginary axis at the origin but curves round again to cut the real axis at $-2/\tau$. It is shown in Figure 18.1.

It is not hard to show that for stability the roots of the characteristic equation must lie inside this circle.

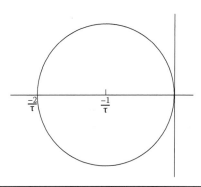

Figure 18.1 β-plane stability circle.

The system

$$\dot{x} = -ax$$

has a pole at $-a$, and is stable no matter how large a is made. On the other hand, the discrete time approximation to this system

$$\beta x = -ax$$

also has a pole at $-a$, but if a is increased to beyond $2/\tau$ the pole emerges from the left margin of the stability circle, and oscillation results. This is the step length problem in another guise.

If we take a set of poles in the s-plane and approximate them by poles in the same position in the β-plane, they must be considerably nearer to the origin than $1/\tau$ for the approximation to ring true. If we want a near-perfect representation, we must map the s-poles to β-poles in somewhat different locations. Then again, the β-plane might not give the clearest impression of what is happening.

18.3 Meet the *z*-Transform

In Chapter 8, we took a quick glance at the z-operator. We needed to use the computer to deduce a velocity signal for controlling the inverted pendulum. Now it is time to consider discrete time theory in more rigorous detail.

The β-operator is a discrete time approximation to differentiation, the counterpart of Euler integration. In Section 2.6 we were able to solve the differential equations to derive an exact simulation that is accurate for any step length. That corresponds to the z-transform.

We saw the Laplace transform defined in terms of an infinite integral

$$\mathcal{L}\big(x(t)\big) = \int_0^\infty x(t)e^{-st}\,dt. \tag{18.3}$$

Now, instead of continuous functions of time, we must deal with sequences of discrete numbers $x(n)$, each relating to a sampled value at time $n\tau$.

We define the z-transform as

$$\mathcal{Z}\big(x(n)\big) = \sum_{n=0}^{\infty} x(n)z^{-n}. \tag{18.4}$$

Even though we know that we can extract $x(r)$ by multiplying the transform by z^r and integrating around a contour that circles $z = 0$, why on earth would we want to complicate the sequence in this way?

The answer is that most of the response functions we meet will be exponentials of one sort or another. If we consider a sampled version of e^{-at} we have

$$x(n) = e^{-ant}$$

so the z-transform is

$$\sum_{n=0}^{\infty} e^{-ant} z^{-n}$$

$$= \sum_{n=0}^{\infty} (e^{-at} z^{-1})^n$$

$$= \frac{1}{1 - e^{-at} z^{-1}}$$

$$= \frac{z}{z - e^{-at}}$$

So we have in fact found our way from an infinite series of numbers to a simple algebraic expression, the ratio of a pole and a zero.

In exactly the same way that a table can be built relating Laplace transforms to the corresponding time responses, a table of sequences can be built for z-transforms. However that is not exactly what we are trying to achieve. We are more likely to use the z-transform to analyze the stability of a discrete time control system, in terms of its roots, rather than to seek a specific response sequence.

A table is still relevant. We are likely to have a system defined by a continuous transfer function, which we must knit together with the z-transform representation of a discrete time controller. But as we will see, the tables equating z and Laplace transforms that are usually published are not the most straightforward way to deal with the problem.

18.4 Trains of Impulses

The z-transform can be related to the Laplace transform, where an *impulse modulator* has been inserted to sample the signal at regular intervals. These samples will take the form of a train of impulses, multiplied by the elements of our number series

$$\sum_{0}^{\infty} x(n)\delta(t - n\tau)$$

we see that its Laplace transform is

$$\int_{0}^{\infty} \left(\sum_{0}^{\infty} x(n)\delta(t - n\tau) \right) e^{-st} dt$$

$$\sum_{0}^{\infty} x(n) \int_{0}^{\infty} \delta(t - n\tau) e^{-st} dt$$

$$\sum_{0}^{\infty} x(n) e^{-ns\tau}$$

If we write z in place of e^{st} this is the same expression as the z-transform.

But to associate z-transforms with inverse Laplace transforms has some serious problems. Consider the case of a simple integrator, $1/s$.

The inverse of the Laplace transform seems easy enough. We know that the resulting time function is zero for all negative t and 1 for all positive t, but what happens at $t = 0$? Should the sampled sequence be

1, 1, 1, 1, 1,....

or

0, 1, 1, 1, 1,....?

Do we sample the waveform "just before" the impulse changes it or just afterwards? Later we will see how to avoid the dilemma caused by impulses and impulse modulators.

Q 18.4.1

What is the z-transform of the impulse-response of the continuous system whose transfer function is $1/(s^2 + 3s + 2)$? (First solve for the time response.) Is it the product of the z-transforms corresponding to $1/(s + 1)$ and $1/(s + 2)$?

18.5 Some Properties of the *z*-Transform

If we have a sequence of which the z-transform is $X(z)$, the transform

$$\frac{1}{z}X(z)$$

$$=\frac{1}{z}\sum_{n=0}^{\infty}x(n)z^{-n}$$

$$=\sum_{n=0}^{\infty}x(n)z^{-n-1}$$

So it will represent the sequence delayed by one sampling interval. Each term occurs one sample later, at time $(n+1)\tau$.

That reinforces our concept of the z-operator as "next."

When we used the beta-transform to approximate

$$\dot{x}=-ax$$

by

$$\beta x = -ax$$

we were really stating that

$$\frac{x(n+1)-x(n)}{\tau}=-ax(n)$$

But we can solve this expression to get a direct value of $x(n+1)$ from $x(n)$:

$$x(n+1)=(1-a\tau)x(n).$$

Knowing the initial value of $x(0)$, we can deduce that

$$x(1)=(1-a\tau)x(0),$$

$$x(2)=(1-a\tau)^2 x(0)$$

and so on until

$$x(n) = (1 - a\tau)^n x(0).$$

Now if $(1 - a\tau)$ has magnitude less than unity, then the magnitude of x will dwindle away as time proceeds. If it is greater than unity, however, the behavior of x will be unstable. Here again is the condition that a must be less than $2/\tau$.

This could possibly simplify first order problems, but any system with a second order or higher derivative will relate to a difference equation involving $x(n + 2)$ or beyond. What do we do then? In the continuous case, we were able to break a high order differential equation down into first order equations. Can we do the same with difference equations? Yes, if we are prepared to define some state variables.

Suppose we have a second order difference equation

$$y(n+2) + 4y(n+1) + 3y(n) = 0 \tag{18.5}$$

so that

$$y(n+2) = -3y(n) - 4y(n+1).$$

Let us define

$$x_1 = y(n),$$

$$x_2 = y(n+1)$$

then we can write two first order difference equations

$$\text{'next'} x_1 = x_2,$$

$$\text{'next'} x_2 = -3x_1 - 4x_2.$$

Before we know it, we have introduced a matrix so that

$$\text{'next'} \begin{bmatrix} x_1 \\ x_2 \end{bmatrix} = \begin{bmatrix} 0 & 1 \\ -3 & -4 \end{bmatrix} \begin{bmatrix} x_1 \\ x_2 \end{bmatrix}$$

and we are back in the world of eigenvectors and eigenvalues. If we can find an eigenvector x with its eigenvalue λ, then it will have the property that

$$\text{'next'}\, x = \lambda x$$

representing stable decay or unstable growth according to whether the magnitude of λ is less than or greater than one.

Note carefully that it is now the magnitude of λ, not its real part, which defines stability. The stable region is not the left half-plane, as before, but is the interior of a circle of unit radius centered on the origin. Only if all the eigenvalues of the system lie within this circle will it be stable. We might just stretch a point and allow the value $(1 + j0)$ to be called stable. With this as an eigenvalue, the system could allow a variable to stand constant without decay.

18.6 Initial and Final Value Theorems

In the case of the Laplace transform we found the initial and final value theorems. The value of the time function at $t = 0$ is given by the limit of $sF(s)$ as s tends to infinity. As s tends to zero, $sF(s)$ tends to the value of the function at infinity.

For the z-transform, the initial value is easy. By letting z tend to infinity, all the terms in the summation vanish except for $x(0)$.

Finding the final value theory is not quite so simple. We can certainly not let z tend to zero, otherwise every term but the first will become infinite.

First we must stipulate that $F(z)$ has no poles on or outside the unit circle, otherwise it will not represent a function which settles to a steady value.

If $G(z)$ is the z-transform of $g(n)$, then if we let z tend to 1 the value of $G(z)$ will tend to the sum of all the elements of $g(n)$. If we can construct $g(n)$ to be the difference between consecutive elements of $f(n)$,

$$g(n) = f(n) - f(n-1)$$

then the sum of the terms of $g(n)$,

$$\sum_{0}^{N} g(n)$$

$$= \sum_{0}^{N} f(n) - f(n-1)$$

$$= f(N) - f(-1)$$

Since $f(-1)$ is zero, as is every negative sample, the limit of this when N is taken to infinity is the final value of $f(n)$. So what is the relationship between $G(z)$ and $F(z)$?

$$G(z) = \sum_{0}^{\infty} (f(n) - f(n-1))z^{-n}$$

$$= \sum_{0}^{\infty} f(n)z^{-n} - \sum_{0}^{\infty} f(n-1)z^{-n}$$

$$= F(z) - \sum_{-1}^{\infty} f(n)z^{-(n+1)}$$

$$= \left(1 - \frac{1}{z}\right)F(z) + f(-1)$$

So we have the final value theory

$$\lim_{n \to \infty} f(n) = \lim_{z \to 1}(1 - z^{-1})F(z) \qquad (18.6)$$

18.7 Dead-Beat Response

Far from being the "poor cousin" of continuous control, discrete time control opens up some extra possibilities. In Section 8.8, we considered a motor system. With feedback, the sequence of values of the position error died away to zero with a time constant of two seconds or so.

In the continuous case, when we allowed all the states to appear as outputs for feedback purposes, we found that we could put the poles of the closed loop system wherever we wished. What is the corresponding consequence in discrete time? In many cases we can actually locate the poles at zero, meaning that the system will settle in a finite number of samples. This is not settling as we have met it in continuous time, with exponential decays that never quite reach zero. This is settling in the grand style, with all states reaching an exact zero and remaining there from that sample onwards. The response is *dead beat*.

Take the example of Q 8.7.1, but now with samples being taken of motor velocity as well as position. When we examined the problem numerically we unearthed a torrent of digits. Let us take more general values for the coefficients, and say that the state equations 8.4 have become

$$\mathbf{x}(n+1) = \begin{bmatrix} 1 & p \\ 0 & q \end{bmatrix} \mathbf{x}(n) + \begin{bmatrix} r \\ s \end{bmatrix} u(n)$$

where the components of **x** are position and velocity. (Here *s* is just a constant, nothing to do with Laplace!) Now we have both states to feed back, so we can make

$$u(n) = [a \quad b] \, \mathbf{x}(n)$$

leading to a closed loop state equation

$$\mathbf{x}(n+1) = \begin{bmatrix} 1+ar & p+br \\ as & p+bs \end{bmatrix} \mathbf{x}(n)$$

The eigenvalues are given by

$$\lambda^{2} - (1 + ar + p + bs) + (p + bp + arp - asp) = 0$$

What is to stop us choosing *a* and *b* such that both these coefficients are zero? Nothing. We can assign values that make both eigenvalues zero by making

$$(1 + p) + ar + bs = 0$$

and

$$p + ap(r - s) + bs = 0.$$

The resulting closed loop matrix is of the form

$$\begin{bmatrix} c & d \\ -c^{2}/d & -c \end{bmatrix}$$

Its rank is only one, and when multiplied by itself it gives the zero matrix—try filling in the details.

In this particular case, the state will settle neatly to zero in the continuous sense too. That need not always be the case, particularly when the system has a damped oscillatory tendency at a frequency near some multiple of the sampling frequency. All that is assured is that the values read at the sampling instants will be zero, and care should always be taken to check the continuous equations for the possibility of trouble in between.

Notice that the designer is faced with an interesting choice. A dead-beat system has a transition matrix with rank less than the order of the system. The square of this matrix has an even lower rank, and at each multiplication the rank must reduce by at least one if it is ultimately to reach zero. Thus a third-order dead-beat system must settle in at most three sampling intervals. Within the limitations of drive amplifiers, a certain settling time is reasonable for a typical disturbance and any faster performance will be dominated by amplifier saturation.

The designer must decide whether to shoot for a dead-beat response that could therefore require a low sampling rate, or whether to sample rapidly and be satisfied with an exponential style of response. Of course saturation need be no bad thing, but the design must then take it into account. The presence or absence of noise will help in making the decision. Also, to achieve a dead-beat performance the designer must have an accurate knowledge of the open-loop parameters. Frequent correction is usually safest, unless the system contains transport delays.

Returning to our example, we have seen the use of observers in the continuous case for constructing "missing" feedback variables. What is the equivalent in discrete time?

18.8 Discrete Time Observers

Although in Section 8.9 we estimated a velocity so that we could control the inverted pendulum experiment, we should take a more formal look at observers in the discrete time case.

In continuous time we made observers with the aid of integrators; now we will try to achieve a similar result with time delays—perhaps with a few lines of software. First consider the equivalent of the *Kalman Filter*, in which the observer contains a complete model of the system. The physical system is

$$\mathbf{x}(n+1) = \mathbf{M}\mathbf{x}(n) + \mathbf{N}\mathbf{u}(n)$$

$$\mathbf{y}(n) = \mathbf{C}\mathbf{x}(n)$$

while the model is

$$\hat{\mathbf{x}}(n+1) = \mathbf{M}\hat{\mathbf{x}}(n) + \mathbf{N}\mathbf{u}(n) + \mathbf{K}(\mathbf{y}(n) - \hat{\mathbf{y}}(n))$$
$$\hat{\mathbf{y}}(n) = \mathbf{C}\hat{\mathbf{x}}(n)$$
(18.7)

Just as in the continuous case, there is a matrix equation to express the discrepancy between the estimated states, here it is a difference equation ruled by the matrix $(\mathbf{M} + \mathbf{KC})$. If the system is observable we can choose coefficients of $\mathbf{K}$ to place the eigenvalues of $(\mathbf{M} + \mathbf{KC})$ wherever we wish. In particular, we should be able to make the estimation dead-beat, so that after two or three intervals an error-free estimate is obtained.

We would also like to look closely at reduced-state observers, perhaps to find the equivalent of phase advance. Now we set up a system within the controller which has state w, where

$$\mathbf{w}(n+1) = \mathbf{P}\mathbf{w}(n) + \mathbf{Q}\mathbf{u}(n) + \mathbf{R}\mathbf{y}(n) \tag{18.8}$$

Now we would like $\mathbf{w}(n)$ to approach some mixture of the system states, $\mathbf{S}\mathbf{x}(n)$. We look at the equations describing the future difference between these values, obtaining

$$\mathbf{w}(n+1) - \mathbf{S}\mathbf{x}(n+1) = \mathbf{P}\mathbf{w}(n) + \mathbf{Q}\mathbf{u}(n) + \mathbf{R}\mathbf{y}(n) - \mathbf{S}\mathbf{M}\mathbf{x}(n) - \mathbf{S}\mathbf{N}\mathbf{u}(n),$$

$$= \mathbf{P}\mathbf{w}(n) + (\mathbf{R}\mathbf{C} - \mathbf{S}\mathbf{M})\mathbf{x}(n) + (\mathbf{Q} - \mathbf{S}\mathbf{N})\mathbf{u}(n). \tag{18.9}$$

If we can bend the equations into the form:

$$\mathbf{w}(n+1) - \mathbf{S}\mathbf{x}(n+1) = \mathbf{P}\{\mathbf{w}(n) - \mathbf{S}\mathbf{x}(n)\}, \tag{18.10}$$

where P describes a system whose signals decay to zero, then we will have succeeded. To achieve this, the right-hand side of Equation 18.9 must reduce to that of 18.10 requiring

$$\mathbf{R}\mathbf{C} - \mathbf{S}\mathbf{M} = \mathbf{P}\mathbf{S}$$

and

$$\mathbf{Q} - \mathbf{S}\mathbf{N} = \mathbf{0} \tag{18.11}$$

Let us tie the design of observer and closed loop response together in one simple example.

By modifying the example of motor-position control so that the motor is undamped, we make the engineering task more challenging and at the same time make the arithmetic very much simpler. To some extent this repeats the analysis we made in Section 8.9, now treated in a rather more formal way.

Q 18.8.1

The motor of a position control system satisfies the continuous differential equation $\ddot{y} = u$. The loop is to be closed by a computer that samples the position and outputs a drive signal at regular intervals. Design an appropriate control algorithm to achieve settling within one second.

First we work out the discrete time equations. This can be done from first principles, without calling on matrix techniques. The state variables are position and velocity. The change in velocity between samples is $u\tau$, so velocity

$$x_2(n+1) = x_2(n) + \tau u(n).$$

Integrating the velocity equation gives

$$x_1(n+1) = x_1(n) + \tau x_2(n) + \tau^2/2\, u(n)$$

so we have

$$\mathbf{x}(n+1) = \begin{bmatrix} 1 & \tau \\ 0 & 1 \end{bmatrix} \mathbf{x}(n) + \begin{bmatrix} \tau^2/2 \\ \tau \end{bmatrix} u(n) \qquad (18.12)$$

Let us first try simple feedback, making $u(n) = ky(n)$. Then the closed loop response to a disturbance is governed by

$$\mathbf{x}(n+1) = \begin{bmatrix} 1+k\tau^2/2 & \tau \\ k\tau & 1 \end{bmatrix} \mathbf{x}(n)$$

The eigenvalues are the roots of

$$\lambda^2 - (2+k\tau^2/2)\lambda + 1 - k\tau^2/2 = 0.$$

If k is positive, the sum of the roots (the negative of the coefficient of λ) is greater than two; if k is negative, the product of the roots (the constant term) is greater than one; we cannot win. The controller must therefore contain some dynamics.

If we had the velocity available for feedback, we could set

$$u = ax_1 + bx_2.$$

That would give us a closed loop matrix

$$\mathbf{x}(n+1) = \begin{bmatrix} 1+a\tau^2/2 & \tau+b\tau^2/2 \\ a\tau & 1+b\tau \end{bmatrix} \mathbf{x}(n)$$

If both eigenvalues are to be zero, both coefficients of the characteristic equation have to be zero. So we need

$$1 + a\tau^2/2 + 1 + b\tau = 0$$

and

$$(1 + a\tau^2/2)(1 + b\tau) - a\tau(\tau + b\tau^2/2) = 0$$

from which we find

$$a\tau^2 = -1$$

and

$$b\tau = -1.5$$

so

$$u(n) = -\frac{x_1(n) + 1.5\tau x_2(n)}{\tau^2} \qquad (18.13)$$

So where are we going to find a velocity signal? From a dead-beat observer. Set up a first order system with one state variable $w(n)$, governed by

$$w(n+1) = p\,w(n) + q\,y(n) + r\,u(n).$$

Equations 18.11 tell us that the state will estimate $x_1(n) + k\,x_2(n)$ if

$$r\mathbf{C} - [1 \quad k]\mathbf{M} = p[1 \quad k]$$

and

$$q - [1 \quad k]\mathbf{N} = 0$$

where $\mathbf{M}$ and $\mathbf{N}$ are the state matrices of Equation 18.12.

If we are going to try for dead-beat estimation, p will be zero. We now have

$$[r \quad 0] - [1 \quad \tau + k] = [0 \quad 0]$$

and

$$q - (\tau^2/2 + k\tau) = 0.$$

From the first of these, $k = -\tau$ and $r = 1$. Substituting in the second, $q = -\tau^2/2$. Now if we make

$$w(n+1) = -\tau^2/2 u(n) + y(n) \tag{18.14}$$

then after one interval w will have value $x_1 - \tau x_2$.

For the feedback requirements of Equation 18.13 we need a mixture in the ratio of $x_1 + 1.5\tau x_2$ to feed back, which we can get from $1.5w - 2.5x_1$. So now

$$u(n) = \frac{3w(n) - 5y(n)}{2\tau^2} \tag{18.15}$$

Now we can re-examine this result in terms of its discrete-transfer function. The observer of Equation 18.14 can be written in transform form as

$$zW = -\tau^2/2U + Y) \tag{18.16}$$

or

$$W = (-\tau^2/2U + Y)/z.$$

This can be substituted into the equation for the controller

$$U = (3W - 5Y)/2\tau^2 \tag{18.17}$$

to give

$$U = -\frac{10 - 6/z}{(4 + 3/z)\tau^2} Y \tag{18.18}$$

This transfer function does in the discrete case what phase advance does for the continuous system. It is interesting to look at its unit response, the result of applying an input sequence (1, 0, 0, …). Let us assume that $\tau = 1$. Equation 18.18 can rearranged as a "recipe" for u in terms of its last value u/z and the present and previous values of y.

$$U = -.75\frac{U}{z} - \frac{10 - 6/z}{4}Y$$

We arrive at the sequence

$$-2.5, 3.375, -2.53125, 1.8984, -1.4238, 1.0679,...$$

After the initial transient, each term is −0.75 times the previous one. This sequence is illustrated in Figure 18.2

Having designed the controller, how do we implement it?

Let us apply control at intervals of 0.5 seconds, so that $\tau^2 = 0.25$, and let us use the symbol w for the observer state in the code. Within the control software, the variable w must be updated from measurements of the output, y, and from the previous drive value u. First it is used as in Equation 18.17 for computing the new drive:

```
u=6*w-10*y;
```

then it is updated by assigning

```
w=y-0.125*u;
```

Now U must be output to the digital-to-analogue converter, and a process must be primed to perform the next correction after a delay of 0.5 seconds. So where have the dynamics of the controller vanished to?

The second line assigns a value to the "next" value of w, zw. This value is remembered until the next cycle, and it is this memory that forms the dynamic effect.

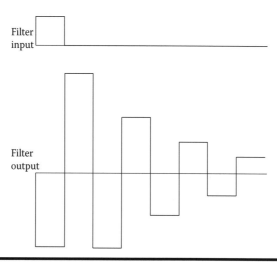

Figure 18.2 Response of the controller.

To test the effectiveness of the controller, it is not hard to simulate the entire system—in this very simple case we use "intuitive" variables x and v.

Let us start with all states zero, including the controller, and then apply a disturbance to the motor position. First we evaluate some coefficients of the discrete-state equations to get:

```
dt=0.5;
A[0]=[1, dt];
A[1]=[0, 1];
B=[dt*dt/2, dt];
```

Now we disturb and mark the initial position

```
x=[1, 0];
LineStart(t, x[0]);
```

and enter the simulation loop, stepping the simulation time:

```
while(t<=5){
   y=x[0];
   u=(3*w-5*y)/(2*dt*dt);
   w=y-u*dt*dt/2;
```

that deals with the controller. Now for the motor simulation:

```
   xnew[0] = A[0][0]*x[0] + A[0][1]*x[1] + B[0]*u;
   xnew[1] = A[1][1]*x[2] + B[1]*u;
   x = xnew;//both vector components are copied
   t=t+dt;
   LineTo(t, x[0]);
  }
```

(The introduction of *xnew* is not really necessary in this particular case.) That completes the program.

Q 18.8.2

Add the necessary details to the code above to construct a "jollies" simulation. Run the simulation for a variety of values of *dt*. Now alter the motor parameter from the assumed value (one unit per second per second) by changing the expressions for *B* to:

```
k=1.2;
B=[k*dt*dt/2, k*dt];
```

Try various values for k. How much can k vary from unity before the performance becomes unacceptable?

A solution is given at www.esscont.com/18/deadbeat.htm

Chapter 19

Relationship between *z*- and Other Transforms

19.1 Introduction

When we come to the use the z-transform in practice, we may have to start from a system that has been defined in terms of its Laplace transform. We must therefore find ways to combine the two forms.

The z-transform involves multiplying an infinite sequence of samples by a function of z and the sample number and summing the resulting series to infinity. The Laplace transform involves multiplying a continuous function of time by a function of s and time and integrating the result to infinity. They are obviously closely related, and in this chapter we examine ways of finding one from the other. We also look at another transform, the w-transform, which is useful as a design method for approximating discrete time control starting from an s-plane specification.

19.2 The Impulse Modulator

In moving between continuous and discrete time in the realm of the Laplace transform, we have had to introduce an artificial conceptual device which has no real physical counterpart. This is the *impulse modulator*. At each sample time, the value of the continuous input signal is sampled and an output is defined which is an impulse having area equal to the value of the sample. Of course this trouble arises because the time function which has a Laplace transform of unity is the unit impulse at $t=0$.

The Laplace transform is concerned with integrations over time, and any such integral involving pulses of zero width will give zero contribution unless the pulses are of infinite height. Could these pulses be made broader and finite? A "genuine" discrete controller is likely to use a *zero order hold*, which latches the input signal and gives a constant output equal to its value, not changing until the next sample time. As we will see, although this device is fundamental to computer control it introduces dynamic problems of its own.

The unit impulse and its transform were discussed in some detail in Sections 14.4 and 15.1. Some other important properties of z-transforms and impulses were considered in Section 18.4.

19.3 Cascading Transforms

When one continuous-time filter is followed by another, the Laplace transfer function of the combination is the product of the two individual transfer functions. A lag $1/(s + a)$ following a lag $1/(s+b)$ has transfer function $1/((s+a)(s+b))$. Can we similarly multiply z-transform transfer functions to represent cascaded elements? When we come to look at the impulse responses, we see that the response of the first filter is e^{-at}. To derive the corresponding z-transform, we must sample this time function at regular intervals τ, multiply the samples by corresponding powers of $1/z$ and sum to infinity.

But we at once meet the dilemma about whether the first sample should be zero or unity. To resolve it, we have to think in engineering terms. At the time of each control activity, first the variables are read and then the output is performed. The first sample has to be zero.

$$Z(e^{-at}) = 0 + z^{-1}e^{-a\tau} + z^{-2}e^{-2a\tau} + \cdots$$

$$= \frac{z^{-1}e^{-a\tau}}{1 - z^{-1}e^{-a\tau}}$$

$$= \frac{e^{-a\tau}}{z - e^{-a\tau}}$$

(19.1)

To construct the z-transform that corresponds to the Laplace transfer function $1/(s+a)(s+b)$ we must first find the function's impulse response. We first split the transfer function into partial fractions

$$\frac{1}{(s + a)(s + b)} = \frac{1}{b - a}\left(\frac{1}{s + a} - \frac{1}{s + b}\right)$$

So the impulse response is

$$\frac{1}{b - a}\left(e^{-at} - e^{-bt}\right)$$

and from Equation 19.1 we see that the z-transform is thus

$$\frac{1}{b-a}\left(\frac{e^{-a\tau}}{z-e^{-a\tau}} - \frac{e^{-b\tau}}{z-e^{-b\tau}}\right)$$

$$\frac{e^{-a\tau}-e^{-b\tau}}{b-a} \quad \frac{z}{(z-e^{-a\tau})(z-e^{-b\tau})} \tag{19.2}$$

On the other hand the product of the two first-order transfer functions is

$$\frac{e^{-(a+b)\tau}}{(z-e^{-a\tau})(z-e^{-b\tau})}$$

Which is clearly different.

When we look at the corresponding waveforms we can see why. The combined system is shown in Figure 19.1 while the two cascaded systems are shown in Figure 19.2.

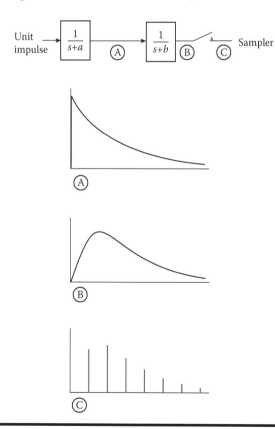

Figure 19.1   Waveforms in the second-order system.

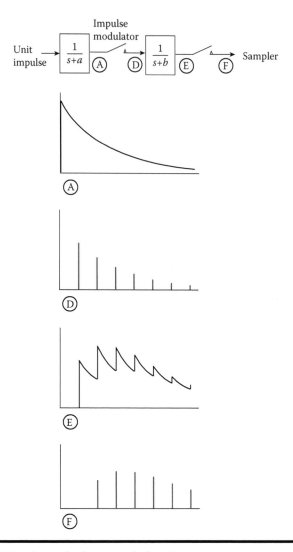

Figure 19.2   Waveforms in the cascaded systems.

When an impulse is applied to the first lag, the response is a decaying exponential in continuous time, or a sequence of decaying samples in discrete time. If a sampling device converts the sequence of values into a corresponding train of impulses and applies them to a second lag, then a second sampler will indeed find a signal as described by the product of the two z-transforms.

The first important lesson is that when systems are connected in cascade, their z-transform cannot in general be found by multiplying the separate z-transforms—unless the subsystems are linked by an impulse modulator. Instead the time-solution of the combined impulse response must be found, and the z-transform calculated by summing the sampled series.

When examining continuous systems, we chose to consider outputs $\mathbf{y}=\mathbf{Cx}$ that were a mixture of the states, leaving out those systems where the outputs could be changed immediately with a change of input. This was a way of avoiding the problem of "algebraic" feedback, where there is an instantaneous loop from the input, through the system and back around the feedback path to the input. The actual value taken up by the input signal is then determined by simultaneous equations.

In well-ordered computer control, this sort of feedback just cannot happen. A signal is sampled, a value is computed and output, and the order of the sampling and computation operations dictates whether one value can influence the other.

19.4 Tables of Transforms

So far we have the following correspondences in Table 19.1.

We will soon need the transform corresponding to $1/s^2$. Is there an easy way to extend the table?

Remember that $n=t/\tau$. If we consider the z-transform of the time function $f(t)$ we have

$$X(z) = Z(f(t))$$

$$= \sum f(n\tau)z^{-n}$$

so $$\frac{d}{dz}X(z) = \sum -nf(n\tau)z^{-n-1}$$

Table 19.1 Transforms and Time Functions

Laplace Transform	Impulse Response	z-Transform
$1/s$	1	$1/(z-1)$
$1/(s+a)$	e^{-at}	$e^{-a\tau}/(z-e^{-a\tau})$
$\dfrac{1}{(s+a)(s+b)}$	$\dfrac{1}{b-a}\left(e^{-at}-e^{-bt}\right)$	$\dfrac{e^{-a\tau}-e^{-b\tau}}{b-a}\dfrac{z}{(z-e^{-a\tau})(z-e^{-b\tau})}$

$$= -\frac{z^{-1}}{\tau} \sum n\tau f(n\tau)z^{-n}$$

$$= -z^{-1}/\tau Z(t\, f(t))$$

In other words,

$$Z(tf(t)) = -\tau z \frac{d}{dz} Z(f(t)). \tag{19.3}$$

The impulse response of $1/s^2$ is t, so the z-transform can be found by differentiating the transform of the constant 1,

$$Z(t) = -\tau z \frac{d}{dz} \frac{1}{z-1}$$

$$= -\tau z \frac{-1}{(z-1)^2} \tag{19.4}$$

$$= \tau \frac{z}{(z-1)^2}$$

19.5 The Beta and *w*-Transforms

We met the beta operator in the previous chapter. As an approximation to differentiation it involved taking the "next" value of the variable it operated on minus its present value, and dividing the result by the time between samples. It could be expressed in terms of the z-transform as

$$\beta = \frac{z-1}{\tau}$$

But we saw that it was unsafe to represent poles in the s-plane by the same poles in the β-plane unless they were close to the origin.

When we look at the relationship between the z-plane delay z^{-1} and the corresponding Laplace delay $e^{-s\tau}$ and regard them as a sort of mapping

$$z = e^{-s\tau}$$

we see that a horizontal "stripe" of the *s*-plane maps onto the entire *z*-plane, as shown in Figure 19.3. Indeed these horizontal stripes map onto the *z*-plane over and over again. If the sample time is τ we will get the same *z*-value from $s + 2n\pi/\tau$ as we do from just *s*.

So how does this come about? When we sample a sinusoidal signal at regular intervals, we encounter *aliasing*. As seen in Figure 19.4, the samples can take the form of a sine-wave in the frequency range $-\pi/\tau$ to $+\pi/\tau$, even when the source frequency is much greater. It is the same sort of effect that causes the spokes of wagon wheels to appear to slow down or rotate backward in old westerns.

We can introduce a new transform that also approximates to differentiation, based on *w* where

$$w = \frac{2}{\tau}\frac{z-1}{z+1}$$

Recall that *z* corresponded to $e^{s\tau}$. Now *w* will correspond to

$$\frac{2}{\tau}\tanh\left(\frac{s\tau}{2}\right)$$

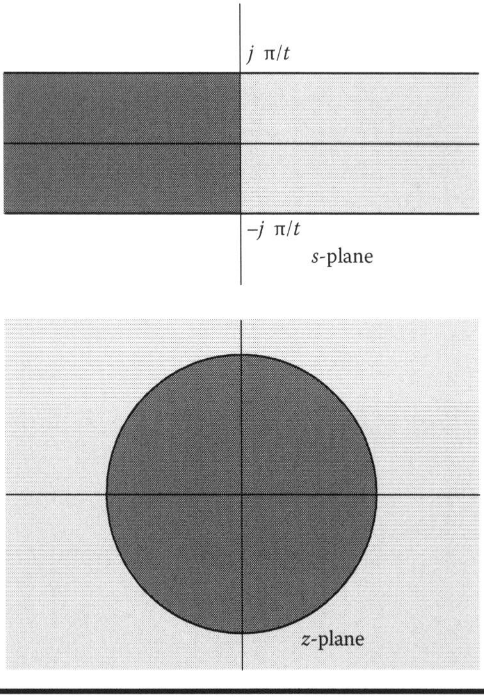

Figure 19.3   Mappings from the *s*-plane to the *z*-plane.

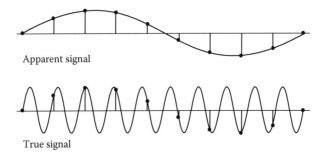

Apparent signal

True signal

Figure 19.4   The effect of aliasing.

where tanh is the *hyperbolic tangent* function. Real values of s now map to real values of w, although the values are "warped." As s tends to minus infinity, w approaches $-2/\tau$.

Since

$$\tanh\left(\frac{j\omega\tau}{2}\right) = j\tan\left(\frac{\omega\tau}{2}\right)$$

we see in Figure 19.5 that the imaginary axis in the s-plane maps to the imaginary axis in the w-plane. But as the frequency approaches π/τ or $-\pi/\tau$, w will dash off toward a positive or negative imaginary infinity.

So what is the use and significance of this transform?

Unlike the β-transform, where the stable region was bounded by a circle, the regions in the w-plane look very much like those in the s-plane. Provided appropriate warping is applied, a discrete time system with rapid sampling can be analyzed in very much the same way used for continuous time. If we know of a compensator that will work in the requency plane, we can find an approximate equivalent for discrete time control.

So how do we account for the improvement over the β-transform?

If we consider an integrator, for example, and write the w-operator as an approximation to differentiation, then

$$W(Y) = U$$

so we have

$$\frac{2}{\tau}\frac{z-1}{z+1}Y = U$$

i.e.,

$$(z-1)Y = \frac{z+1}{2}\tau U$$

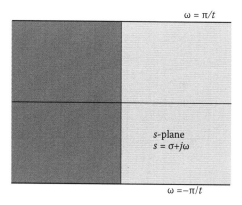

As w tends to j^* infinity, ω tends to π/t

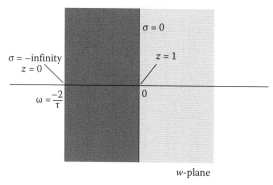

Figure 19.5   Mapping between the z and w-planes.

Instead of the change in y being proportional to the present input, it is proportional to the average of this and the next values of input. In effect it corresponds to *trapezoidal* integration rather than the simpler *Euler integral* represented by the β-transform. These are compared in Figures 19.6 and 19.7.

Its need to depend on the next value of input makes y unsuitable to be a state variable. But the output y can be constructed from a conventional summing variable with an amount $\tau/2$ of the input added to it.

```
x = x + tau*u
y = x + tau*u/2
```

So if the sampling rate is high compared with the system time constants, poles can be placed in the w-plane that approximately correspond to the poles in the s-plane. The poles and zeros of a continuous controller that is known to work can be represented by poles in the w-plane.

Now these w-plane poles can be converted to z-plane poles using

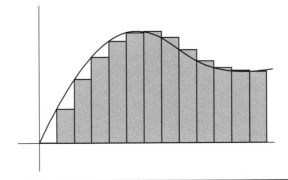

Figure 19.6 Rectangular Euler integration. The integral is the area under the blocks.

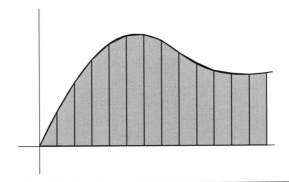

Figure 19.7 Trapezoidal integration.

$$w = \frac{2}{\tau} \frac{z-1}{z+1}$$

from which

$$z = \frac{2+w\tau}{2-w\tau}$$

and finally the z-plane poles and zeros of the controller enable the controller code to be written.

Q 19.5.1

In a system sampled 10 times per second, what discrete time controller corresponds to the phase advance $(1+2s)/(2+s)$?

Chapter 20

Design Methods for Computer Control

20.1 Introduction

In Chapter 18, we saw that the state-space methods of continuous time could be put to good use in a discrete time control system. Now we look at further techniques involving transforms and the frequency domain. Some design methods seem straightforward, such as the root locus, while others can conceal pitfalls. When continuous and discrete feedback are mixed, analysis can be particularly difficult.

20.2 The Digital-to-Analog Convertor (DAC) as Zero Order Hold

The conventional presentation of z-transforms involves the unit impulse, that awkward function of infinite magnitude and zero width. But in any practical digital control system, the result of the computation is a simple finite number. It is output to the real world as a simple value that stays constant between one iteration and the next.

Sometimes the value is converted to a mark-space switching waveform, but in many cases the translation circuit is a *digital-to-analog convertor*. The nature of the convertor depends very much on the application. At the simplest level it may consist of no more than a relay or valve that delivers power to a heater or actuator. Another application may require the output to be a current accurate to one part in many thousands, perhaps to deflect an electron beam in a silicon fabrication process. For now we will look at the substantial middle ground, where a conversion accuracy of one part in 256 is more than sufficient.

If you rely on a conventional table of Laplace transforms, several stages are involved in finding the z-transform that relates the controller's output to the samples taken from the system being controlled.

Suppose that we have a sequence of outputs 1, 4, 8, 5, 3, 2, 6...., as shown in Figure 20.1.

Each time a new value is output, the DAC makes a step change. The changes are the differences between the latest value of output and the previous one. You will note that we cannot specify the value of the first such step without knowing the DAC level before it. This change is ? 3, 4, –3, –2, –1, 4....

To get the steps, we can regard them as the output of an integrator driven by a sequence of impulses with these values.

The business of taking the difference between this output value and the previous one is very simply performed by the digital filter

$$1 - \frac{1}{z}$$

But then we have to perform an integration as a continuous operation and that integrator has to be considered as part of the continuous system.

Another way of looking at the task is to recognize that we need to know the z-transform of the output sample sequence when the input has a z-transform of just 1. This input sequence is 1, 0, 0, 0,… with a waveform shown in Figure 20.2.

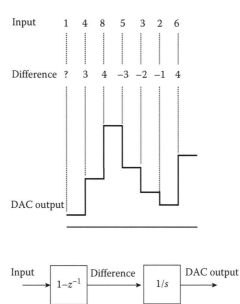

Figure 20.1 DAC output steps are the differences of the input values.

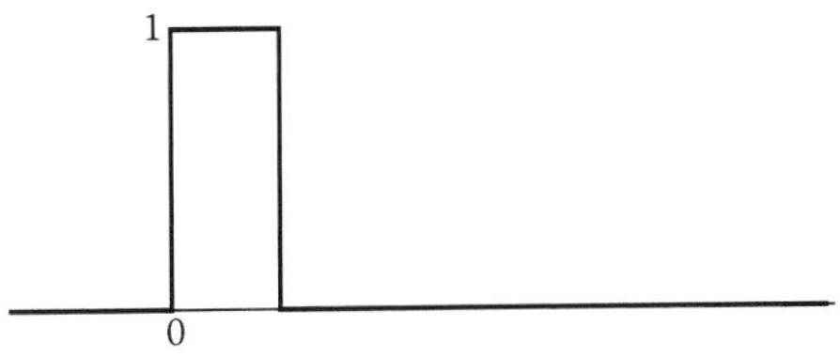

Figure 20.2 Control waveform with a z-transform of 1.

To construct it we must add a unit step at $t = 0$ and subtract another one at the next sample time.

Now when we consider transfer functions, we should look at a time response to this function instead of the response to a unit impulse.

Bearing in mind that the new input is applied after the sample has been taken, we have a transform for a unit transfer function that is $1/z$, a unit delay.

An integrator $1/s$ will ramp up to the value τ during the first interval, then stay constant, giving a sample sequence

$$0, \tau, \tau, \tau, \ldots$$

and thus have a transform

$$\frac{\tau}{z - 1}$$

We will find a table of z-transforms of *pulse responses*, rather than impulse responses as in Table 19.1, much more convenient to use.

Q 20.2.1

What is the z-transform equivalent of a motor described by $s^2 Y = U$, when driven by a DAC updated 10 times per second? Can the motor be controlled stably by position feedback alone?

Q 20.2.2

Can the motor position be stabilized using the "digital phase advance" of Section 8.9?

20.3 Quantization

In passing it is worth mentioning quantization errors.

If a controller outputs a position demand to an analog position control subsystem, the command might require a great number of levels to define the target with sufficient precision. On the other hand, if the digital output commands a motor drive there is little point in seeking great accuracy, since it will be surrounded by a tight position feedback loop.

Eight-bit DAC's are common. They are of low cost, since the manufacturing tolerance is easy to meet. Most control computers are equipped to output eight-bit bytes of data, so there is little incentive to try to use a simpler DAC.

The *quantization* of a DAC means that a compromise has been made to round the output value to the nearest DAC value available. When 256 levels define all the output values from fully positive to full negative, each step represents nearly one percent of the amplitude from zero. The effect is of a "noise" waveform being added to the signal, with an amplitude of half of one DAC bit.

20.4 A Position Control Example, Discrete Time Root Locus

The analog part of the system of Q 20.2.1 is described by

$$Y(s) = \frac{1}{s^2}U(s) \tag{20.1}$$

We have several ways of looking up the equivalent z-transform in a table of transform equivalents, but it is more instructive to work it out ourselves. The impulse response is the time function

$$y(t) = t^2/2.$$

Using the formula of Equation 19.3, we can differentiate the transform of the function t to get

$$Y(z) = -\tau z \frac{d}{dz} \frac{\tau z}{(z-1)^2}$$

$$= \frac{\tau^2}{2} \frac{z(z+1)}{(z-1)^3}$$

for the impulse response. For the transform of the unit pulse response we must multiply by $(1 - z^{-1})$ to get

$$Y(z) = -\frac{\tau^2}{2} \frac{(z+1)}{(z-1)^2}. \tag{20.2}$$

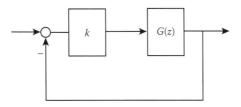

Figure 20.3   Unity feedback around a variable gain.

We can obtain the same result by calculating the transfer function matrix from the matrix state equations, as we did in Section 18.8.

Now if we close the loop, using a gain k at the input as shown in Figure 20.3, we have a closed loop gain

$$\frac{kG(z)}{1+kG(z)}$$

We can look at the poles algebraically by multiplying out the denominator, obtaining

$$(z-1)^2 + k\frac{\tau^2}{2}(z+1)$$

i.e., the equation for the poles is

$$z^2 + z\left(-2+k\frac{\tau^2}{2}\right)+\left(1+k\frac{\tau^2}{2}\right)=0$$

As we found in Section 18.8, neither positive nor negative values of k can give stability.

But maybe we can get more insight from a root locus, since we have a variable gain k to explore.

We have two poles at $z=1$ and a zero at $z=-1$. The rules at the end of Sections 12.4 and 12.5 give us the following clues for sketching the root locus:

(1) For large k, a single "excess pole" makes off to infinity along the negative real axis.
(2) Only that part of the real axis to the left of the zero can form part of the plot for positive k. (i.e., for negative feedback.)
(3) The derivative of $G(z)$ is zero at $z=-3$, a point that therefore lies on the locus. (Exercise for the reader). This point is therefore a junction of branches of the locus.

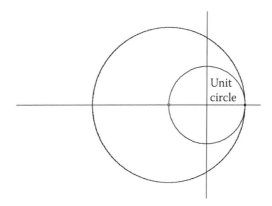

Figure 20.4 Root locus for proportional feedback. (Screen grab of www.esscont. com/20/rootlocz.htm)

We could deduce that the locus splits and leaves the pole pair "North and South," following a pair of curved paths that come together again at $z = -3$. One branch now follows the real axis to the zero at $z = -1$, while the other travels along the axis in the opposite direction to minus infinity. The curved part of the locus is in fact a circle, and the locus is illustrated in Figure 20.4, generated with much less effort by the software at www.esscont.com/20/rootlocz.htm. It is plain that the locus lies entirely in the forbidden region outside the unit circle, so that stability is impossible.

By adding a dynamic controller with extra poles and zeros in the loop, can we "bend" the locus to achieve stability?

20.5 Discrete Time Dynamic Control–Assessing Performance

Can we stabilize the position system with the digital equivalent of a phase advance? Suppose we take the transfer function $(1 + 2s)/(2 + s)$ that was proposed in example Q 19.5.1. We use the w-transform to find an approximate equivalent. First we substitute w for s, where

$$w = \frac{2}{\tau} \frac{z-1}{z+1}$$

to get a transfer function,

$$\frac{1+2w}{2+w}$$

$$= \frac{1+2\dfrac{2(z-1)}{\tau(z+1)}}{2+\dfrac{2(z-1)}{\tau(z+1)}}$$

$$= \frac{(\tau+4)z+(\tau-4)}{(2\tau+2)z+(2\tau-2)}$$

$$= \frac{4.1z-3.9}{2.2z-1.8}$$

when we substitute the value of 0.1 for τ.

A zero at $z = 0.951$ and a pole at $z = 0.818$ are added to the root locus diagram, with a result illustrated in the plots shown in Figure 20.5. Stable control is now possible, but the settling time is not short. How can we tell?

In the negative-real half of the s-plane, the real part of s defines an exponential decay rate while the imaginary part defines a frequency. A complex pair of roots can be viewed as the roots of a quadratic

$$s^2 + 2\zeta\omega_0 s + \omega_0^2 = 0$$

where ω_0 is the undamped natural frequency and ζ is the damping factor.

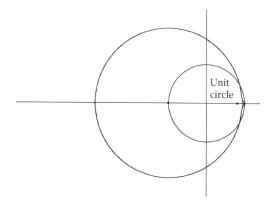

Figure 20.5 **Root locus with compensator. (Screen grab of www.esscont.com/20/rootzcomp.htm)**

The damping factor dictates the shape of the step response while the natural frequency determines its speed. A collection of responses for varying values of ζ are shown in Figure 20.6.

Now constant values of ω_0 will correspond to points on circles around the origin of the *s*-plane. Constant values of damping factor will lie on lines through the origin at an angle $\cos^{-1}\zeta$ to the negative real axis. This is shown in Figure 20.7.

A value of $\zeta = 0.7$ gives a response of acceptable appearance. This is a very popular choice of damping factor, apparently for no more complicated reason than that the poles lie on lines drawn at 45° through the *s*-plane origin! The natural

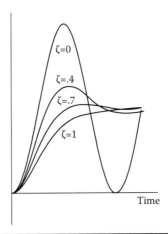

Figure 20.6 Responses for a variety of damping factors.

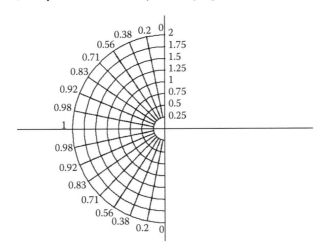

Figure 20.7 Lines of constant damping factor and of constant undamped natural frequency.

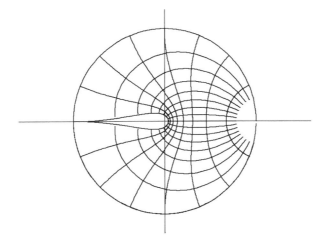

Figure 20.8   *z*-plane loci for constant ζ and for constant ω_0.

frequency, the frequency at which the system would oscillate if the damping were removed, is now represented by the distance from the origin.

To translate these yardsticks into the discrete time world, we must use the relationship $z = e^{s\tau}$. Damping factors and natural frequencies are mapped into the interior of the unit circle as shown in Figure 20.8.

Now we are in a position to judge how effective a compensator can be by examining its root locus. We can clearly do better than the gentle phase advance we started with. We could perhaps mix the feedback signal with a pseudo-derivative formed by taking the signal and subtracting its previous value

$$u = a\, y(n) + b\,(y(n) - y(n-1))$$

i.e.,

$$U = ((a+b) - b/z)Y.$$

This gives a pole at the origin and a zero that we can place anywhere on the segment of the real axis joining the origin to $z = 1$. A root locus for the case where the zero is at $z = 0.5$ is shown in Figure 20.9. In software it would be implemented for some chosen value of gain by

```
u=-k*(2*y-yold);
yold=y;
```

Would the performance of the system have been more easily deduced from its *s*-plane transfer function, using the approximate equivalence of *s* to *w*? If we

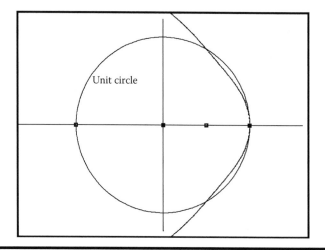

Figure 20.9 Root locus with "differencing" controller, a zero at 0.5 and pole at the origin.

substitute w for s, and then replace w by $2(z - 1)/\tau(z + 1)$ we will have a system transfer function with three zeros at $z = 1$ and three poles at $z = -1$. Adding in the "differencer" removes one of the poles, but adds another pole at $z = 0$. The result bears little resemblance to the system we must actually control. The w-transform may enable us to devise filters, but could be misleading for modeling a system.

On the other hand, the z-transform transfer function can be translated exactly into the w-plane by substituting $(2 + w\tau)/(2 - w\tau)$ for z. Now a w-plane compensator can be tried out, plotting the resulting root locus in the w-plane. The stability criterion is the familiar one of requiring all the poles to have negative real parts.

Q 20.5.1

Represent the motor position problem by a transfer function in w; sketch its root locus for proportional feedback. What is the effect of a compensator with transfer function $(1 + w)$? Can such a compensator be realized?

Making the substitution $z = (2 + w\tau)/(2 - w\tau)$ in Equation 20.2 gives us a transfer function $(1 - w\tau/2)/w^2$ for the DAC and motor combination. This has two poles at the origin and a zero at $w = 20$. The root locus illustrated in Figure 20.10 shows that feedback alone will not suffice.

The compensator $(1 + w)$ adds a second zero at $w = -1$. Do the twin poles now split, one moving to the left and the other to the right along the real axis to each zero?

Not a bit.

The transfer function has an extra negative sign, seen when it is rewritten as $-(w\tau/2 - 1)/w^2$, so the valid real-axis part of the locus lies outside the zeros,

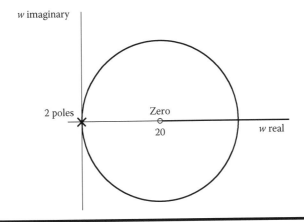

Figure 20.10 ■ *w*-plane root locus for uncompensated feedback.

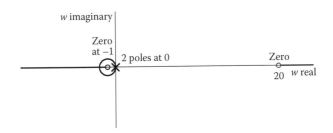

Figure 20.11 ■ *w*-plane root locus with *(1 + w)* compensation.

not between them. The poles split North and South, following a circular path to recombine on the negative real axis. One pole now approaches the zero at $w = -1$, while the other makes off to minus infinity, reappearing at plus infinity to return along the real axis to the second zero. Figure 20.11 shows the result.

The compensator is not hard to realize. Another substitution, this time replacing $(1 + w)$ by $(1 + 2(z - 1)/0.1(z + 1)$, gives us the transfer function

$$k\frac{21z - 19}{z + 1}$$

which can be rewritten as

$$k\frac{21 - 19z^{-1}}{1 + z^{-1}}$$

So we have

$$U = -z^{-1}U + k(21 - 19z^{-1})Y$$

The code to calculate this is

```
u=-uold+k*(21*y-19*yold);
uold=u;
yold=y;
```

When preceded by an input statement to read y from the sampler and followed by a command to output u to the DAC, this will construct the compensator in question. On its own the compensator looks marginally unstable, but when incorporated in the loop it should be able to perform as predicted.

It appears deceptively simple to place a few compensator poles and zeros to achieve a desired response. There must be further objectives and criteria involved in designing a control system. Indeed there are. Our problems are just beginning.

But first let us look at a final root locus. In Sections 18.7 and 18.8 a controller was devised to give a dead-beat response. Expression 18.18 shows that it has a pole at a negative real value of z, at $z = -0.75$, giving an even more dramatic modification of the root locus. The zero is now at $z = 0.6$, and the result is illustrated in Figure 20.12. Note that all the roots come together at the origin.

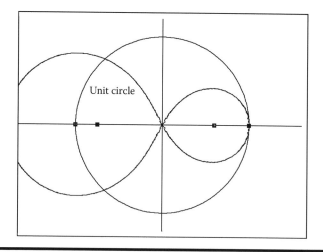

Figure 20.12   Root locus giving a dead-beat system.

Chapter 21

Errors and Noise

21.1 Disturbances

Until now, our systems have included inputs, states, and outputs. Our disturbances have been represented by initial conditions, and we have assumed that the goal of the controller is to bring the error to zero in the absence of any other input. This is an oversimplification, since the objective of many controllers is to combat disturbances that afflict the system.

In the water-heater experiment of Chapter 5, we saw the PID controller proposed as a way to correct for a standing error. It is a useful device that industry applies to a variety of problems. With the derivative (or phase advance) action of the "D" term added to the proportional "P" term, many second-order systems can be brought to stability. The integral "I" term means that the drive will continue to increase until the error has been driven to zero—or until the limit of maximum drive is reached.

Many principles of controller design have arisen from the needs of gunnery. A controller containing one integration can reduce to zero the aiming error with a stationary target, but will have a tracking error proportional to the target's velocity. With a second integrator we can track a target without error at constant velocity, but the aim will be thrown off by acceleration. Each additional integrator raises the order of the polynomial that can be followed, but introduces a more prolonged transient behavior in acquiring the target. Fortunately most industrial problems can be solved with no more than a single integration.

Integral control is a method open to a simple analog controller. When the controller is digital, many more sophisticated techniques are available. The controller can be "aware" of a change of demand, and integral correction can pause until the response has settled. The error can then be integrated over a finite, predefined interval of time, to find the average correction needed and this can be applied as a step change of input. After a delay to allow the transient to pass, the integration can be repeated. In this way, there is a hope of achieving dead-beat correction of the offset, instead of the slow exponential decay resulting from continuous integral feedback.

Now we must investigate the various forms of disturbance in a more general way with the aid of some block diagrams as shown in Figure 21.1.

N1: The command input is subject to noise of various forms, from quantization of a numeric command to tolerances in encoding a control lever. This error is outside the loop and therefore nothing can be done about it. The target will be the signal as interpreted by the control electronics.

N2: An error in measuring the value of the output for feedback purposes is equally impossible to compensate. It is the sensor signal that the loop corrects, rather than the true output. If a position transducer has slipped 10°, then the controller will steadfastly control the output to a position with a 10° error.

N3: This represents the quantization error of the feedback digitizer. While N2 arises by accident; N3 is deliberately permitted by the system designer when selecting the fineness of the digitization, perhaps in terms of the bit-length of an analog-to-digital convertor (ADC).

N4: The computation of the algorithm will not be perfect, and extra disturbances can be introduced through rounding errors, through shortened multiplications and approximated trigonometric functions.

N5: Another portion of quantization error is served up when driving the digital to analog convertor (DAC) output. The magnitude of this error is determined by the system designer when selecting the DAC word-length.

N6: The DAC is a fallible electronic device, and unless carefully set up can be subject to offset error, to gain error and to bit-sensitive irregularities.

N7: This is the noise disturbing the system that the controller has really been built to combat. It can range from turbulent buffeting in an aircraft to the passenger

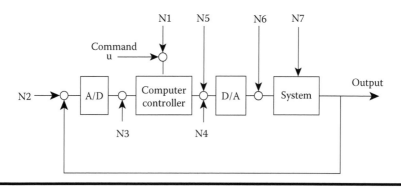

Figure 21.1   Block diagram of a system with noise sources.

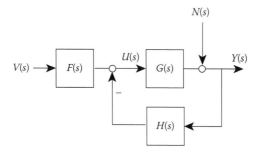

Figure 21.2   Control loop with disturbance noise.

load on an escalator, from the switching of an additional load onto a stabilized power supply to the opening of a refrigerator door.

In a linear system, the noise source N7 can be replaced by a transformed signal N(s) representing its effect at the output of the system. The equations of the feedback system shown in Figure 21.2 become

$$Y(s) = G(s)U(s) + N(s),$$

$$U(s) = F(s)V(s) + H(s)Y(s).$$

Now

$$Y(s) = G(s)\{F(s)V(s) - H(S)Y(s)\} + N(s),$$

$$\{1 + G(s)H(s)\}Y(s) = G(s)F(s)V(s) + N(s)$$

or

$$Y(s) = \frac{G(s)F(s)}{1 + G(s)H(s)}V(s) + \frac{1}{1 + G(s)H(s)}N(s)$$

Our aim must be to minimize the effect of the disturbance, $N(s)$. If $V(s)$ is a command representing the desired value of the output, then we also want the first transfer function to be as near to unity as possible. In the tracking situation, the system may be simplified and redrawn as shown in Figure 21.3.

To establish the ability of the system to track perturbations of various kinds, we can substitute functions for $N(s)$ representing a step, a ramp, or a higher power of time. Then we can use the final-value theorem to find the ultimate value of the error.

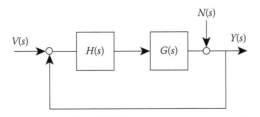

Figure 21.3   A tracking system with noise.

Q 21.1.1

Show that with an $H(s)$ of unity the system $1/s(s+1)$ can track a step without a final error, but that $1/(s+1)$ cannot.

Q 21.1.2

Show that by adding integral control, with $H(s) = 1 + a/s$, the system $1/(s+1)$ can be made to track a step with zero final error.

The use of large gains will impose heavy demands on the drive system; for any substantial disturbance the drive will limit. Large derivative or difference terms can also increase the drive requirements, made worse by persistent high frequency noise in the system. If the drive is made to rattle between alternate limiting extremes, the effect is to reduce the positional loop gain so that offset forces will result in unexpectedly large errors.

The pursuit of rapid settling times will have a penalty when noise is present, or when demand changes are likely to be large. Linear design techniques can be used for stabilization of the final output, but many rapid-response positioning systems have a separate algorithm for handling a large change, bringing the system without overshoot to the region where proportional control will be effective. As we have seen, the design of a limiting controller requires the consideration of many other matters than eigenvalues and linear theory.

21.2 Practical Design Considerations

We often find that the input to an ADC is contaminated by the presence of noise. If the noise is of high frequency it may be only necessary to low-pass filter the signal to clean it up. We have already seen that the one-line computation

```
x=x+(signal-x)/k
```

will act as a filter with approximate time constant k times the sampling interval. Will this remove contaminating noise? Unfortunately we are caught out by aliasing.

Whereas an analog low-pass filter will attenuate the signal more and more as the frequency increases, the digital filter has a minimum gain of $1/k$. Indeed as the input frequency increases toward the sampling frequency the gain climbs again to a value of unity, as shown in Figure 21.4.

The frequency response is seen from a frequency plane diagram, see Figure 21.5. A sine-wave is represented by a point that moves around the unit circle. The gain is inversely proportional to the length of the vector joining that point to the pole at $1/k$.

We cannot use digital filtering, but must place an analog low-pass filter before the input of the ADC. This will probably provide a second benefit. An ADC has a limited input range, corresponding to its, say, 256 possible answers. For efficient use of the conversion, the input signal range should nearly "fill" that of the ADC. If allowance has to be made in the range for the presence of noise on top of the signal, a smaller signal amplitude must be used. The quantization noise of then becomes more important.

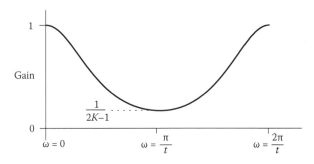

Figure 21.4 Frequency response of a low-pass digital filter.

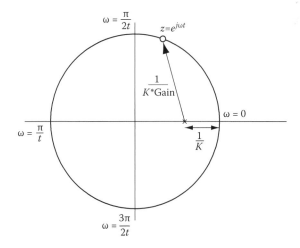

Figure 21.5 The gain is inversely proportional to the vector.

There is, however, a situation where digital filtering can be more effective than analog. Signals in industrial environments, such as those from strain-gauges, are often contaminated with noise at the supply frequency, 50 or 60 Hz. To attenuate the supply frequency with a simple filter calls for a *break frequency* of the order of one second. This may also destroy the information we are looking for. Instead we can synchronize the convertor with the supply frequency, taking two or four (or more) readings per cycle. Now taking the sum of one cycle of readings will cause the supply frequency contribution to cancel completely—although harmonics may be left. With 12 readings per cycle, all harmonics can be removed up to the fifth.

It is possible that the input signal contains data frequencies well above the supply frequency, so that even the filtering suggested above would obscure the required detail. It is possible to sample at a much higher multiple of the supply frequency, and to update an array which builds an average from cycle to cycle of the measurement at each particular point in the cycle. In other words, the array reconstructs the cyclic noise waveform. This can be subtracted from the readings to compensate each sample as soon as it is taken.

Care is needed when selecting an ADC for the feedback signal. It is obvious that the quantization error must be smaller than the required accuracy, but there are other considerations. If the control algorithm depends on differential action of any form, the quantization effect can be aggravated. If sampling is rapid, so that there are many control cycles between changes in value of the digitized signal, then the difference between subsequent samples will appear as sporadic unit pulses as seen in Figure 21.6.

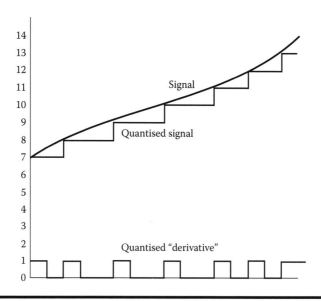

Figure 21.6 A signal is quantized and then differences are taken.

When the open-loop system is not in itself well damped, it often makes the control task much easier if analog feedback is applied around a tight "inner loop." Thus a position controller can have analog velocity feedback, permitting the output of the digital controller to represent a demand for velocity rather than for acceleration.

The analysis of a system with an analog inner loop is relatively straightforward, although some design decisions are now separated and more difficult to make. The continuous system can be expressed in terms of Laplace operators, and root-locus, block diagram or algebraic methods can help decide on the feedback details. This entire block can then transformed back into the time domain and sampled, so that a z-transform representation can be made for the path from DAC at the system input to ADC sensing the output. The command input is also in sampled form, so that it can be represented by a z-transform, and the closed loop transfer function can be calculated as the ratio of the z-transform of the output divided by the z-transform of the input. Now the digital control and compensation can be designed to place the closed loop z poles to the designer's choice.

If the command input arrives in analog form as shown in Figure 21.7, and rather than being sampled by the controller is applied directly to the system, then it is another story. Just as the z-transform of two cascaded analog filters is unlikely to be the product of their individual transforms, so the z-transform of the output will not be the z-transform of the command input multiplied by the z-transform of the system response. If a digital outer loop is applied, say to apply integral control, then the system cannot really be analyzed in z-transform terms. Instead, the digital part of the system will probably be approximated to its continuous equivalent, allowing design to be performed in the s-plane.

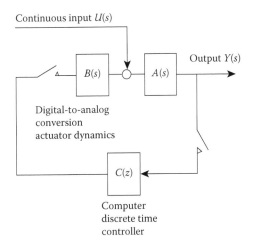

Figure 21.7 It can be impossible to derive a transfer function if a system has both analog and digital feedback.

21.3 Delays and Sample Rates

The choice of sampling rate can be important, particularly when the system contains a pure time delay associated with an otherwise stable system. I am sure that you are familiar with the task of adjusting the temperature of a bathroom shower—the sort that has a length of pipe between the mixer and the shower-head. The successful strategy is to try a setting, then have the patience to wait until the new mixture has passed through the system before making a compensating adjustment. If you are too hasty, you perpetrate a temperature oscillation that leaves you alternately chilled and scalded.

The z-transform of a delay of one sample is of course $1/z$. If the same delay is sampled twice as fast, the transfer function becomes $1/z^2$. Far from improving the digital control problem, an excessive sampling rate stacks up poles at the origin that seriously limit the possible speed of response. With a single such pole, dead-beat control might be a possibility.

The analysis of a delay of two-and-a-half samples can be rather tricky. The transform in terms of the impulse response is zero, since the sampler that builds the output sequence will miss any impulse modulator signals by half a sample-width. In terms of the unit DAC pulse, however, the answer is much more sensible.

For a system with no other dynamics, the sample sequence will be 0, 1, 0, 0,..., since we have stipulated that the sample must be taken before the output is applied. The transfer function will thus be z^{-1}. For any delay up to one sample, the function will be the same. For any delay from one sample to just under two, the function will be z^{-2}. For a delay of two-and-a-half samples the function will thus be z^{-3}.

When there are added dynamics, the situation is more complicated. For an added integrator, the output will ramp from zero to τ during the period that the unit pulse reaches the integrator, which we can assume precedes the delay. The sample sequence will thus be 0, 0, 0, 0, $\tau/2$, τ, τ,..., which will have transform

$$\frac{\tau}{2}z^{-4} + \frac{z^{-5}}{1 - z^{-1}}$$

With a more complicated system preceding or following the delay, solving the pulse response will yield an answer for any length of delay.

The choice of output DAC depends very much on the target performance. On–off control is adequate for an amazing variety of applications, particularly in domestic temperature regulation. However, control of a robot or an autopilot is likely to warrant the extra cost of a DAC of at least eight-bit precision. The accuracy of the DAC does not contribute greatly to the accuracy of the final output position. It is the accuracy of the feedback DC that is important here. It could however make a great difference to the "ride," in terms of vibration and smoothness of control.

In the struggle for increased performance there is a temptation to over-specify the output device, be it servomotor or heater. It is always necessary first to analyze the possible failure modes, which will usually include a "runaway condition." An autopilot whose servomotor is powerful enough to pile on 30° of aileron in a fraction of a second is going to scare the hardiest of pilots, let alone the airline passengers. A fine compromise must be made between the maximum required manoeuvre and the limit of safe failure—the system should "fail soft."

21.4 Conclusion

Practical controller design can involve as much art as science, particularly when there are several inputs and variables to be controlled. In an autopilot it was traditional to use the elevator to control height and the throttle to control airspeed. Suddenly it was realized that it is more logical to alter the glide angle to control the speed, and to add extra thrust and energy to gain height; the roles of the throttle and aileron were reversed. Both control systems worked and maybe had the same poles, but their responses were subtly different.

Advanced analysis methods can aid the designer in the matter of choosing parameters, but they should not restrict the choice of alternative strategies out of hand. Even the selection of one damping factor over another may be as much a matter of taste as of defined performance. With the ability to add non-linearities at will, with the possibility of mixing discrete and continuous controllers that defy analysis, as a system designer you might be tempted to use a rule-of-thumb. Make sure that you first get to grips with the theory and gain familiarity through simulation and practice, so that when the time comes your thumb is well calibrated.

Chapter 22

Optimal Control— Nothing but the Best

22.1 Introduction: The *End Point* Problem

Until now, our analysis assumes that the control will go on forever. But there are many control tasks that have *fixed end points*, or end points in general. Then, instead of fretting about poles and zeros, we have real criteria that have to be met.

- The elevator must approach the floor slowly enough that when the brakes are put on the passengers do not bounce.
- The missile or torpedo must pass close enough to the target that its explosion can destroy it.
- The unmanned lunar module must touch down before running out of fuel.
- The aircraft must touch down with small vertical velocity, with wings nearly level, with small sideslip across the runway, pointing straight down the runway, with enough remaining runway to stop safely.
- We must reach the next gas station before we run out of fuel.

These are not criteria that can be addressed by a transfer function. They require a *cost function*. This can relate to the end point, such as "fuel remaining," or can be something integrated throughout the journey. The simplest example of this is arrival time, where we can integrate a constant 1 and also get time as a state variable.

We have seen that few systems are linear, that our matrix state equations are a mathematical convenience. But there are state equations for all the more general systems, at least for those that possess state equations:

299

$$\dot{\mathbf{x}} = \mathbf{f}(\mathbf{x}, \mathbf{u})$$

Here $\mathbf{f}$ is a vector function with an element for each of the state variables x_1 to x_n. It can contain all the nonlinearities resulting from drive limitation and any interactions between the variables.

For assessing our control, if we have a cost that is an integral we can express the cost function as

$$\int_{\text{start}}^{\text{target}} c(\mathbf{x}, \mathbf{u}) dt$$

But suppose that we define a new state variable:

$$\dot{x}_0 = c(\mathbf{x}, \mathbf{u}) \tag{22.1}$$

then our cost will simply be the final value of x_0. We will have rolled all our tasks into the same form of looking at the state at the end point.

In fact we will have two end points to consider. If we wish to fire a missile at a target flying by we must consider and control the starting time.

For practical purposes the cost criterion can usually be set at "good enough." But for mathematicians, optimal control seems to have a particular attraction. By seeking the optimal solution, they can present us with an answer that is unique.

As soon as the controller develops beyond a simple feedback loop, the system designer is spoilt for choice. Where are the best positions to place the closed loop poles? How much integral control should be applied? These agonizing decisions can be resolved by looking for the "best" choice of coefficients, those that will minimize the chosen cost function.

Though the single mission tasks threaten to require some innovative control techniques, there is a class of cost functions that result in good old faithful linear control. As we will see, this has been the target of considerable theory. The cost function usually involves the integral of the square of the error at any time. To result in a solution that does not ask the drive to be a maximum, some multiple of the square of the drive signal is added in too.

But then instead of pondering poles and zeros, the designer must agonize over the particular mix of the cost function.

22.2 Dynamic Programing

In seeking optimal control, we must use a rather different version of calculus from that involved in solving for differential equations.

In its "pure" form it becomes the *Calculus of Variations*. A classic example of the theory is in looking at the shape of a loose hanging chain, suspended between two points. The chain must sag in such a way that its potential energy is minimized. The approach is to assume that we already know the solution, the function $y(x)$ that determines the shape, and then we work out the effect of a small perturbation. This will be a change in the potential energy, but we have another equation expressing the condition that the length must remain constant.

An intuitive way to wrap our minds around the optimization theory is to look at *dynamic programing*.

Despite an impressive name, the principle of Bellman's Dynamic Programing is really quite straightforward. It simply states that if the system has reached an intermediate point on an optimal path to some goal, then the remainder of the path must be the optimal path from that intermediate point to the goal.

It seems very simple, but its application can lead to some powerful theories. Consider the following example.

The car will not start, your train to the city leaves in 12 minutes time, the station is nearly two miles away. Should you run along the road or take the perhaps muddy short cut across the fields? Should you set off at top speed, or save your wind for a sprint finish?

If you could see all possible futures, you could deduce the earliest possible time at which you might reach the station, given your initial state. Although there would be best strategy that corresponded to this time, the best arrival time would be a function of the state alone.

Given your present fitness, distance from the station, muddiness of the field, and the time on the station clock, you can reach the station with one minute to spare. But every second you stand trying to decide on whether to take the short cut, one second of that minute ticks away.

You make the correct decision and start to sprint along the road. If an all-knowing computer clock could show your best arrival time, then the value displayed would stand still. You put on an extra burst of speed around the bend, but the computer display ticks forward, not back. You are overtiring yourself, and will soon have to slow down.

There is no way that you can pull that best arrival time backward, because it is the optimum, the best that you can achieve in any circumstance. If you pace yourself perfectly, keeping exactly to the best route, you can hold the arrival time steady until you arrive.

Five minutes later you are over half way there, hot, flustered, getting tired, but in sight of the station. You have lost time. The prediction clock shows that you have only 30 seconds to spare. That indicates the best arrival time from your new state. For each second that passes, you have to reduce the time required for the remainder of the journey by one second. If you could but see that prediction display, you could judge whether a burst of effort was called for. If the computed best arrival time moves it can only move forward, the decision is wrong.

Now let us firm up the homespun philosophy into a semblance of mathematical reality.

In this case the cost is going to be the integral of a cost function:

$$\int_{start}^{target} c(\mathbf{x}, \mathbf{u}) dt$$

where in this example the cost function c is the constant, unity, since its integral is the journey time we are trying to minimize.

By defining the new state variable x_0 to be the total cost, as suggested in Equation 22.1, we can just regard the values at the end point.

For any initial state $\mathbf{x}$, there is a best possible cost $C(\mathbf{x})$. You apply an input $\mathbf{u}$, taking you to a new state $\mathbf{x} + \delta\mathbf{x}$ by time $t + \delta t$. Now the best possible cost from your new state will be the function C of the new variables, so your best cost will be

$$C(\mathbf{x} + \delta\mathbf{x}).$$

You can influence this total cost by your choice of input $\mathbf{u}$. For optimal control you will select the value of $\mathbf{u}$ that minimizes the cost. But the best you can do is to hold the value to be the same as the best cost from your starting point, in other words:

$$\min_{\mathbf{u}} \left(C(\mathbf{x} + \delta\mathbf{x}) \right) = C(\mathbf{x})$$

or

$$\min_{\mathbf{u}} \left(C(\mathbf{x} + \delta\mathbf{x}) - C(\mathbf{x}) \right) = 0 \tag{22.2}$$

For the best possible choice of input, the rate change of the best cost will be zero, i.e.,

$$\min_{\mathbf{u}} \left(\frac{d}{dt} C(\mathbf{x}) \right) = 0 \tag{22.3}$$

Here the derivative of C is total, taking into account all the changes of the components of the state, $\mathbf{x}$. If we expand the total derivative in terms of the rates of change of all the states, this becomes

$$\min_{\mathbf{u}} \left(\sum_{i=0,n} \frac{dx_i}{dt} \frac{\partial}{\partial x_i} C(\mathbf{x}) \right) = 0 \tag{22.4}$$

We have two conclusions to reach. First, we must choose the input that will minimize expression 22.4. Secondly, for this best input, the expression in the brackets will become zero

Now we have been weaving ever-increasing webs of algebra around the function *C* without really having a clue about what form it takes. It is an unknown, but not a complete mystery. We are starting to amass information not about the function itself, but about its partial derivatives with respect to the states. Let us define these derivatives as variables in their own rights, and see if we can solve for them. We define

$$p_i = \frac{\partial C}{\partial x_i} \tag{22.5}$$

so that the function we are trying to minimize is then

$$\sum_{i=0,n} \frac{dx_i}{dt} p_i$$

We substitute for $\dot{x}_i$ from our state equations to write it as

$$\sum_{i=0,n} f_i(\mathbf{x}, \mathbf{u}) p_i(\mathbf{x})$$

So let us define a *Hamiltonian*,

$$H = \sum_{i=0,n} f_i p_i \tag{22.6}$$

We are trying to minimize *H* with respect to each of the inputs u_j, so we look at

$$\frac{\partial H}{\partial u_j} = \sum_{i=0,n} \frac{\partial f_i}{\partial u_j} p_i \tag{22.7}$$

since *C*, and hence **p**, is not a function of **u**. Either this will become zero in the allowable range of inputs, representing an acceptable minimum, or else the optimum input variable must take a value at one end or other of its limited range. In either event, when the optimal values of the inputs are substituted back into the expression for *H*, to give the rate of change of total cost, the result must be zero.

Indeed, *H* must be zero both throughout the trajectory and on any neighboring optimal trajectory. Thus its partial derivative with respect to a state variable will also be zero. So

$$\frac{\partial H}{\partial x_j} = \sum_{i=0,n} \left(\frac{\partial f_i}{\partial x_j} p_i + f_i \frac{\partial p_i}{\partial x_j} \right) = 0 \tag{22.8}$$

Maybe we can find out more about **p** by taking a derivative with respect to time. Since **p** is only a function of **x**, its rate of change will be seen in terms of the rates of change of the state variables:

$$\frac{dp_j}{dt} = \sum_{i=0,n} \dot{x}_i \frac{\partial p_j}{dx_i}$$

i.e.,

$$\dot{p}_j = \sum_{i=0,n} f_i \frac{\partial p_j}{dx_i} \tag{22.9}$$

Equation 22.8 tells us that

$$\sum_{i=0,n} \frac{\partial f_i}{\partial x_j} p_i + \sum_{i=0,n} f_i \frac{\partial p_i}{\partial x_j} = 0$$

so

$$\sum_{i=0,n} f_i \frac{\partial p_i}{\partial x_j} = -\sum_{i=0,n} \frac{\partial f_i}{\partial x_j} p_i \tag{22.10}$$

Now from our definition 22.5:

$$\frac{\partial p_j}{dx_i} = \frac{\partial}{dx_i} \frac{\partial C}{\partial x_j} = \frac{\partial^2 C}{\partial x_i \partial x_j}$$

If *C* obeys certain continuity conditions, so that we may take liberties with the order of partial differentiation, this will be equal to

$$\frac{\partial p_i}{\partial x_j}$$

so

$$\dot{p}_j = \sum_{i=0,n} f_i \frac{\partial p_i}{dx_j}$$

and using Equation 22.10 we see that

$$\dot{p}_j = -\sum_{i=0,n} \frac{\partial f_i}{\partial x_j} p_i$$

i.e., by swapping i and j,

$$\dot{p}_i = -\sum_{j=0,n} \frac{\partial f_j}{\partial x_i} p_j \tag{22.11}$$

which we can alternatively write as

$$\dot{p}_i = -\frac{\partial H}{\partial x_i} \tag{22.12}$$

Well this is an impressive looking result, but what does it mean?

22.3 Optimal Control of a Linear System

When we have our familiar matrix state equation:

$$\dot{\mathbf{x}} = \mathbf{A}\mathbf{x} + \mathbf{B}\mathbf{u}$$

we can expand it as

$$\dot{x}_i = \sum a_{ij} x_j + \sum b_i u_i$$

The partial derivative

$$\frac{\partial f_i}{\partial x_j}$$

is thus, equal to a_{ij}.

Equation 22.11 then becomes a matrix equation:

$$\dot{\mathbf{p}} = -\mathbf{A}'\mathbf{p} \tag{22.13}$$

where $\mathbf{A}'$ is the transpose of $\mathbf{A}$.

We have another set of state equations to solve if we want to know the value of **p**. According to *Pontryagin's Maximum Principle*, the vector **p** is the *adjoint vector* and the matrix $-\mathbf{A}'$ is the *adjoint matrix*.

Then we have simply to choose the inputs to minimize the Hamiltonian that is now

$$H = \mathbf{p}'\mathbf{A}\mathbf{x} + \mathbf{p}'\mathbf{B}\mathbf{u} \tag{22.14}$$

Let us see how it works in practice.

22.4 Time Optimal Control of a Second Order System

Let us start gently, with the system $\ddot{y} = u$. The state equations are

$$\dot{x}_1 = x_2$$

$$\dot{x}_2 = u$$

The cost function is elapsed time, represented by x_0 where

$$\dot{x}_0 = 1$$

Now we have

$$H = p_0 f_0 + p_1 f_1 + p_2 f_2$$
$$= p_0 + p_1 x_2 + p_2 u.$$

(22.15)

We must minimize H in terms of u, so u will take a value at one or other limit given by

$$u = -u_{max} \, \text{sgn}(p_2).$$

All we have to do to apply time-optimal control is to solve for p_2. Now the differential equations for the p's are given by the derivatives of H by x_i, see Equation 22.15, and since H only involves x_2 of these:

$$\dot{p}_0 = 0$$

$$\dot{p}_1 = 0$$

$$\dot{p}_2 = -p_1$$

The first of these equations is satisfied by $p_0 = 1$, which is as it should be, while the second and third give

$$p_1 = a,$$

$$p_2 = -at + b.$$

So now we know that the sign of u is given by that of $(at - b)$.

Have we solved the entire problem? These equations hold no more clues about the choice of the constants a and b. The best we can say that the optimal input involves applying full drive with at most one reversal of sign.

This form of analysis and that of *Calculus of Variations* are concerned with the manner of controlling the system, not with its details. A classic application of Calculus of Variations proves that the shortest distance between two points is a straight line. If I ask a villager how to get to the station by the shortest route, and I receive the reply "Walk in a straight line," I am not altogether satisfied.

Nevertheless, the theory has narrowed down the search for an optimal strategy to just two decisions: the initial sign of the drive, and the time at which it must reverse. If the target is to reach zero error at zero velocity, then a strategy unfolds. The last part of the trajectory must bring the state to rest at the origin of the state-space—more graphically seen as the origin of the phase-plane. There will be two full-drive trajectories through the origin, one for each sense of drive. In this undamped case they will be parabolae. They divide the phase-plane into two regions, as shown in Figure 22.1. In one of these the drive must be maximum positive, in the other it must be maximum negative.

We apply full drive toward the origin, putting on the brakes at the last moment from which we can avoid an overshoot.

There are many ways to apply the appropriate drive signal using an analog circuit. The most obvious is the use of a nonlinear function generator. A shaping circuit can compute the error multiplied by its modulus, which is then added to the velocity, applied to a comparator and the resulting sign passed to the drive. Alternatively the position error signal can be added to a shaped velocity signal, representing the sign of the velocity times the square root of its magnitude, and a comparator will give the same result.

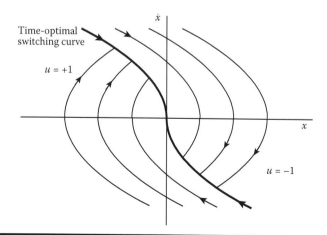

Figure 22.1 Time-optimal switching curve.

22.5 Optimal or Suboptimal?

My own doctorate research stemmed from a particularly interesting technique termed "Predictive Control." It involves the use of a fast simulator model. Devised by John F. Coales and Maxwell Noton, it stimulated additional groups under Harold Chestnut and Boris Kogan to develop simplifications and enhancements. In effect, since there are few decisions to be made a fast model can set out from the present state, can be subjected to a trial input and be interrogated about the consequences. From the result, a new fast-model input is determined, and a decision is made about the input to the process.

Chestnut's second order strategy is particularly simple. In the model, braking drive is applied to bring the velocity to zero. If the model stops with a negative error then positive plant drive must be applied, and vice versa.

I considered higher order systems. For a three-integrator system, the drive may have two switches of sign. For n integrators it may have $n - 1$ switches. Predictive strategies to compute optimal control will rapidly become complicated, as interaction between the effects of the individual switching times is more complex and less predictable. One proposal involved models within models, each one faster than the last.

Is optimal control really necessary, or even desirable? It can be shown via the Maximum Principle that the fuel-optimal soft lunar landing is achieved by burning the engines at full rate until touchdown. This involves igniting the engines at the last possible moment. When the spaceship approaches the moon before it decelerates, it will be traveling in the region of one mile per second. If ignition is one second late, the result is disaster.

The fuel usage is related closely to the momentum that must be removed, which is being continually increased at the rate of one lunar gravity all the time the spaceship is descending. If the motors fire earlier, and continue under slightly reduced thrust, then the landing will be slightly delayed and the momentum increased. But the increase in use of fuel for a really substantial safety margin of thrust is extremely small—and well worth the investment. For a non-return vehicle, any fuel that remains has no value whatsoever.

But to return to predictive control. A strategy for systems consisting of three cascaded integrators involved running the model with each sign of drive but with no switches. The plant drive was determined in terms of which trajectory was "offside" furthest ahead in model time. "Offside" was defined to mean that the position error and the model drive were of opposite signs. This strategy was investigated by one of my students, Song Feng Mo.

When brushing up some software for a new student joining me a year ago, I made a slip in the software and an amazing discovery. If "offside" is taken to mean that *any* of the model states have the opposite sign to its drive, then the strategy works well for four and five integrators—and who knows how many more. You can investigate the strategy yourself at www.esscont.com/22/predictive.htm.

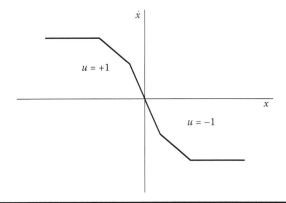

Figure 22.2   Piecewise linear suboptimal controller.

With the ready availability of low-cost embeddable computer power, the day for fast-model predictive strategies may have arrived.

For a position controller, time-optimal control of Figure 22.1 might not give the most desirable response. At the origin, the switching line between positive and negative drive is vertical, and the input will buzz between extremes while the position dithers slightly. It can be beneficial to the lifetime of gears and motors to replace the inner section of the switching curve with a linear approximation of limited slope, see Figure 22.2, and even to reduce the gain of the comparator from infinity to a mere few thousands.

Even when we seem to be returning to a linear system, the linear system design guidelines are inappropriate. The slope of the position/velocity switching line defines the time constant of a first order type response, an exponential decay with a time-constant of, say, 0.1 seconds. There may be a second real root at one millisecond or less, as the loop gain is tightened, which suggests a damping factor of five or more:

$$s^2 + 1010s + 10000 = 0.$$

Any temptation to reduce the velocity term should be firmly resisted. The high loop gain arises not through a desire for faster response, but out of a need for precision. The constant term of 10,000 indicates that full drive will be applied if the position error exceeds 0.01% of one unit.

22.6 Quadratic Cost Functions

Before leaving optimal control, let us look again at the general problem of optimal control of a linear system.

We saw that if we followed the Maximum Principle, our drive decisions rested on solving the adjoint equations:

$$\dot{\mathbf{p}} = -\mathbf{A}'\mathbf{p} \tag{22.16}$$

For every eigenvalue of $\mathbf{A}$ in the stable left-hand half-plane, $-\mathbf{A}'$ has one in the unstable half-plane. Solving the adjoint equations in forward time will be difficult, to say the least. Methods have been suggested in which the system equations are run in forward time, against a memorized adjoint trajectory, and the adjoint equations are then run in reverse time against a memorized state trajectory. The *boresight* method allows the twin trajectories to be "massaged" until they eventually satisfy the boundary conditions at each end of the problem.

When the cost function involves a term in $\mathbf{u}$ of second order or higher power, there can be a solution that does not require bang-bang control. The quadratic cost function is popular in that its optimization gives a linear controller. By going back to dynamic programing, we can find a solution without resorting to adjoint variables, although all is still not plain sailing.

Suppose we choose a cost function involving sums of squares of combinations of states, added to sums of squares of mixtures of inputs. We can exploit matrix algebra to express this mess more neatly as the sum of two *quadratic forms*:

$$c(\mathbf{x}, \mathbf{u}) = \mathbf{x}'\mathbf{Q}\mathbf{x} + \mathbf{u}'\mathbf{R}\mathbf{u}. \tag{22.17}$$

When multiplied out, each term above gives the required sum of squares and cross-products. The diagonal elements of $\mathbf{R}$ give multiples of squares of the u's, while the other elements define products of pairs of inputs. Without any loss of generality, $\mathbf{Q}$ and $\mathbf{R}$ can be chosen to be symmetric.

A more important property we must insist on, if we hope for proportional control, is that $\mathbf{R}$ is *positive definite*. The implication is that any nonzero combination of u's will give a positive value to the quadratic form. Its value will quadruple if the u's are doubled, and so the inputs are deterred from becoming excessively large. A consequence of this is that $\mathbf{R}$ is non-singular, so that it has an inverse.

For the choice of $\mathbf{Q}$, we need only insist that it is positive semi-definite, that is to say no combination of x's can make it negative, although many combinations may make the quadratic form zero.

Having set the scene, we might start to search for a combination of inputs which would minimize the Hamiltonian, now written as

$$H = \mathbf{x}'\mathbf{Q}\mathbf{x} + \mathbf{u}'\mathbf{R}\mathbf{u} + \mathbf{p}'\mathbf{A}\mathbf{x} + \mathbf{p}'\mathbf{B}\mathbf{u}. \tag{22.18}$$

That would give us a solution in terms of the adjoint variables, $\mathbf{p}$, which we would still be left to find. Instead let us try to estimate the function $C(\mathbf{x}, t)$ that expresses the minimum possible cost, starting with the expanded criterion:

$$\min_{\mathbf{u}} \left(c(\mathbf{x}, \mathbf{u}) + \frac{\partial C}{\partial t} + \sum_{i=1,n} \frac{\partial C(\mathbf{x},t)}{\partial x_i} \dot{x}_i \right) = 0 \qquad (22.19)$$

If the control is linear and if we start with all the initial state variables doubled, then throughout the resulting trajectory both the variables and the inputs will also be doubled. The cost clocked up by the quadratic cost function will therefore, be quadrupled. We may, without much risk of being wrong, guess that the "best cost" function must be of the form:

$$C(\mathbf{x}, t) = \mathbf{x}' \mathbf{P}(t) \mathbf{x}. \qquad (22.20)$$

If the end point of the integration is in the infinite future, it does not matter when we start the experiment, so we can assume that the matrix $\mathbf{P}$ is a constant. If there is some fixed end-time, however, so that the time of starting affects the best total cost, then $\mathbf{P}$ will be a function of time, $\mathbf{P}(t)$.

So the minimization becomes

$$\min_{\mathbf{u}} \left(\mathbf{x}' \mathbf{Q} \mathbf{x} + \mathbf{u}' \mathbf{R} \mathbf{u} + \mathbf{x}' \dot{\mathbf{P}} \mathbf{x} + \sum_{i=1,n} 2(\mathbf{x}' \mathbf{P})_i \dot{x}_i \right) = 0$$

i.e.,

$$\min_{\mathbf{u}} \left(\mathbf{x}' \mathbf{Q} \mathbf{x} + \mathbf{u}' \mathbf{R} \mathbf{u} + \mathbf{x}' \dot{\mathbf{P}} \mathbf{x} + 2\mathbf{x}' \mathbf{P} (\mathbf{A} \mathbf{x} + \mathbf{B} \mathbf{u}) \right) = 0 \qquad (22.21)$$

To look for a minimum of this with respect to the inputs, we must differentiate with respect to each u and equate the expression to zero.

For each input u_i,

$$2(\mathbf{R} \mathbf{u})_i + 2(x' \mathbf{P} \mathbf{B})_i = 0$$

from which we can deduce that

$$\mathbf{u} = -\mathbf{R}^{-1} \mathbf{B}' \mathbf{P} \mathbf{x}. \qquad (22.22)$$

It is a clear example of proportional feedback, but we must still put a value to the matrix, $\mathbf{P}$. When we substitute for $\mathbf{u}$ back into Equation 22.21 we must get the answer zero. When simplified, this gives

$$x'(\mathbf{Q} + \mathbf{P} \mathbf{B} \mathbf{R}^{-1} \mathbf{B}' \mathbf{P} + \dot{\mathbf{P}} + 2\mathbf{P} \mathbf{A} - 2\mathbf{P} \mathbf{B} \mathbf{R}^{-1} \mathbf{B}' \mathbf{P}) x = 0$$

This must be true for all states, x, and so we can equate the resulting quadratic to zero term by term. It is less effort to make sure that the matrix in the brackets is symmetric, and then to equate the whole matrix to the zero matrix. If we split $2\mathbf{PA}$ into the symmetric form $\mathbf{PA} + \mathbf{A'P}$, (equivalent for quadratic form purposes), we have

$$\dot{\mathbf{P}} + \mathbf{PA} + \mathbf{A'P} + \mathbf{Q} - \mathbf{PBR}^{-1}\mathbf{B'P} = 0.$$

This is the matrix *Riccati* equation, and much effort has been spent in its systematic solution. In the infinite-time case, where $\mathbf{P}$ is constant, the quadratic equation in its elements can be solved with a little labor.

Is this effort all worthwhile? We can apply proportional feedback, where with only a little effort we choose the locations of the closed loop poles. These locations may be arbitrary, so we seek some justification for their choice. Now we can choose a quadratic cost function and deduce the feedback that will minimize it. But this cost function may itself be arbitrary, and its selection will almost certainly be influenced by whether it will give "reasonable" closed loop poles!

Q 22.6.1

Find the feedback that will minimize the integral of $y^2 + a^2u^2$ in the system $\dot{y} = u$.

Q 22.6.2

Find the feedback that will minimize the integral of $y^2 + b^2\dot{y}^2 + a^2u^2$ in the system $\ddot{y} = u$.

Before reading the solutions that follow, try the examples yourself. The first problem is extremely simple, but demonstrates the working of the theory. In the matrix state equations and quadratic cost functions, the matrices reduce to a size of one-by-one, where

$$\mathbf{A} = 0,$$

$$\mathbf{B} = 1,$$

$$\mathbf{Q} = 1,$$

$$\mathbf{R} = a^2, \quad \text{so} \quad \mathbf{R}^{-1} = 1/a^2.$$

Now there is no time-limit specified, therefore, $d\mathbf{P}/dt = 0$.
We then have the equation:

$$\mathbf{PA} + \mathbf{A'P} + \mathbf{Q} - \mathbf{PBR}^{-1}\mathbf{B'P} = 0$$

to solve for the "matrix" **P**, here just a one-by-one element p.

Substituting, we have

$$0+0+1- p.1.(1/a^2).1.p = 0$$

i.e.,

$$p^2 = a^2$$

Now the input is given by

$$u = -\mathbf{R}^{-1}\mathbf{B}'\mathbf{P}\,y$$
$$= -(1/a^2).1.ay$$
$$= -y/a$$

and we see the relationship between the cost function and the resulting linear feedback.

The second example is a little less trivial, involving a second order case. We now have two-by-two matrices to deal with, and taking symmetry into account we are likely to end up with three simultaneous equations as we equate the components of a matrix to zero.

Now if we take y and $\dot{y}$ as state variables we have

$$\mathbf{A} = \begin{bmatrix} 0 & 1 \\ 0 & 0 \end{bmatrix}$$

$$\mathbf{B} = \begin{bmatrix} 0 \\ 1 \end{bmatrix}$$

$$\mathbf{Q} = \begin{bmatrix} 1 & 0 \\ 0 & b^2 \end{bmatrix}$$

$$\mathbf{R} = a^2$$

The matrix P will be symmetric, so we can write

$$\mathbf{P} = \begin{bmatrix} p & q \\ q & r \end{bmatrix}$$

Once again $d\mathbf{P}/dt$ will be zero, so we must solve

$$\mathbf{PA} + \mathbf{A'P} + \mathbf{Q} - \mathbf{PBR}^{-1}\mathbf{B'P} = 0$$

so

$$\begin{bmatrix} 0 & p \\ 0 & q \end{bmatrix} + \begin{bmatrix} 0 & 0 \\ p & q \end{bmatrix} + \begin{bmatrix} 1 & 0 \\ 0 & b \end{bmatrix} - \begin{bmatrix} p & q \\ q & r \end{bmatrix} \begin{bmatrix} 0 \\ 1 \end{bmatrix} \frac{1}{a^2} \begin{bmatrix} 0 & 1 \end{bmatrix} \begin{bmatrix} p & q \\ q & r \end{bmatrix} = \begin{bmatrix} 0 & 0 \\ 0 & 0 \end{bmatrix}$$

i.e.,

$$\begin{bmatrix} 1 & p \\ p & b+2q \end{bmatrix} - \frac{1}{a^2}\begin{bmatrix} q^2 & qr \\ qr & r^2 \end{bmatrix} = \begin{bmatrix} 0 & 0 \\ 0 & 0 \end{bmatrix}$$

from which we deduce that

$$q^2 = a_2,$$

$$qr = a^2 p$$

and

$$r^2 = a^2(b+2q)$$

from which $q = a$ (the positive root applies), so $r = a\sqrt{2a+b}$ and $p = \sqrt{2a+b}$. Now u is given by

$$u = -\mathbf{R}^{-1}\mathbf{B'Px},$$

$$u = -\frac{1}{a^2}\begin{bmatrix} 0 & 1 \end{bmatrix}\begin{bmatrix} \sqrt{2a+b} & a \\ a & a\sqrt{2a+b} \end{bmatrix}\begin{bmatrix} y \\ \dot{y} \end{bmatrix}$$

$$u = -\frac{1}{a}y - \frac{\sqrt{2a+b}}{a}\dot{y}$$

It seems a lot of work to obtain a simple result. There is one very interesting conclusion, though. Suppose that we are concerned only with the position error and do not mind large velocities, so that the term b in the cost function is zero. Now our cost function is simply given by the integral of the square of error plus a multiple of the square of the drive. When we substitute the equation for the drive into the system equation, we see that the closed loop behavior becomes

$$\ddot{y} + \sqrt{2}\sqrt{\frac{1}{a}}\dot{y} + \frac{1}{a}y = 0$$

Perhaps there is a practical argument for placing closed loop poles to give a damping factor of 0.707 after all.

22.7 In Conclusion

Control theory exists as a fundamental necessity if we are to devise ways of persuading dynamic systems to do what we want them to. By searching for state variables, we can set up equations with which to simulate the system's behavior with and without control. By applying a battery of mathematical tools we can devise controllers that will meet a variety of objectives, and some of them will actually work. Other will spring from high mathematical ideals, seeking to extract every last ounce of performance from the system, and might neglect the fact that a motor cannot reach infinite speed or that a computer cannot give an instant result.

Care should be taken before putting a control scheme into practice. Once the strategy has been fossilized into hardware, changes can become expensive. You should be particularly wary of believing that a simulation's success is evidence that a strategy will work, especially when both strategy and simulation are digital:

"A digital simulation of a digital controller will perform exactly as you expect it will—however catastrophic the control may be when applied to the real world."

You should by now have a sense of familiarity with many aspects of control theory, especially in the foundations in time and frequency domain and in methods of designing and analyzing linear systems and controllers. Many other topics have not been touched here: systems identification, optimization of stochastic systems, and model reference controllers are just a start. The subject is capable of enormous variety, while a single technique can appear in a host of different mathematical guises.

To become proficient at control system design, nothing can improve on practice. Algebraic exercises are not enough; your experimental controllers should be realized in hardware if possible. Examine the time responses, the stiffness to external disturbance, the robustness to changing parameter values. Then read more of the wide variety of books on general theory and special topics.

Index